TRAINEE GUIDE

FOR

PREFLIGHT

C-9B-0020

Unit 1

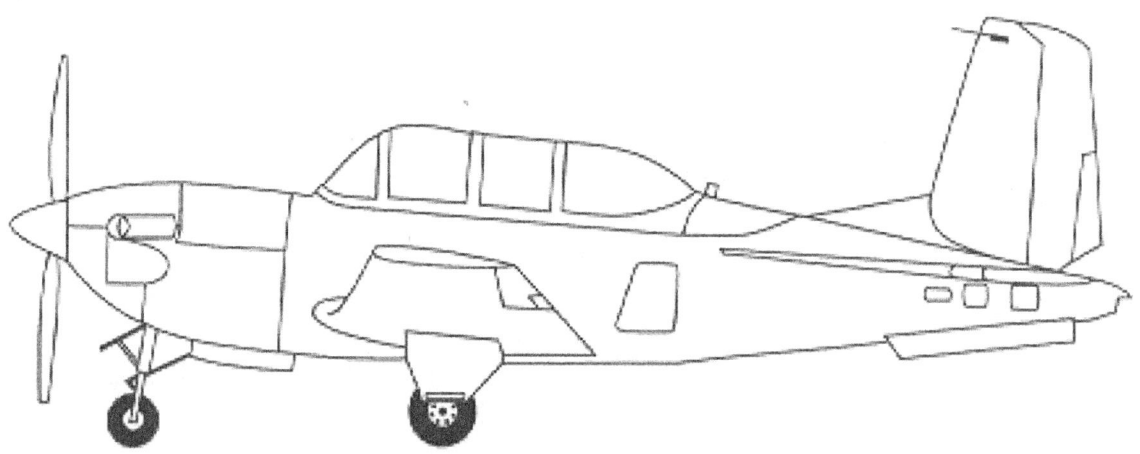

FUNDAMENTALS
OF
AERODYNAMICS

Prepared by

NAVAL AVIATION SCHOOLS COMMAND

181 CHAMBERS AVE SUITE C

PENSACOLA, FL 32508

Prepared for

CENTER FOR NAVAL AVIATION TECHNICAL TRAINING

230 Chevalier Field Ave Suite C

Pensacola, FL 32508

April 2008

CHANGE RECORD

Number	Description of Change	Entered by	Date

TABLE OF CONTENTS

SECURITY AWARENESS NOTICE

This course does not contain any classified material.

SAFETY NOTICE

All personnel must be reminded that personal injury, death or equipment damage can result from carelessness, failure to comply with approved procedures, or violations of warnings, cautions, and safety regulations.

SAFETY / HAZARD AWARENESS NOTICE

Safe training is the number one goal. Each year at training commands, lives are lost and thousands of man hours and millions of dollars are wasted as the result of accidents. Most of these accidents could have been prevented. They are the result of actions performed incorrectly, either knowingly or unknowingly, by people who fail to exercise sufficient foresight, lack the requisite training, knowledge, or motivation, or who fail to recognize and report hazards.

A mishap is any unplanned or unexpected event causing personnel injury, occupational illness, death, material loss or damage or an explosion whether damage occurs or not.

A near miss or hazardous condition is any situation where if allowed to go unchecked or uncorrected has the potential to cause a mishap.

It is the responsibility of all Department of Defense personnel to report all mishaps and near misses. If a mishap, hazardous condition or near miss occurs let your instructor know immediately.

Students will report all hazardous conditions and near misses to the command high-risk safety officer via their divisional/departmental high-risk safety officer. Reports can be hand written on the appropriate form. Injuries shall be reported on the appropriate form.

HOW TO USE THIS STUDENT GUIDE

This publication is for your use while studying aerodynamics. It is designed to be specific to the T-34C. It will also provide you with a basic foundation which you will build upon during training in more advanced aircraft. You may mark any pages in this book, including information sheets and assignment sheets. When filled in, this guide will become a useful reference. You may not use it during testing.

The aerodynamics unit is presented in two sections. The first section is divided into Lesson Topics 1 through 6 covering basic principles of physics, aircraft terminology, basic aerodynamic principles, lift, drag, thrust and power. The second section is divided into Lesson Topics 7 through 12 covering more advanced topics such as airplane performance, controls, stability, spins, turning flight, and takeoff and landing performance.

The knowledge to be acquired is stated for each topic so that you can check your progress. It is to your advantage to review the learning objectives prior to the class presentation.

Assignments in this guide are given for study. The effectiveness of the guide depends upon the conscientious accomplishment of the reading and study assignments.

Participation in a study group is highly recommended. Statistical analysis suggests that a study group of four members is optimum.

A written examination will be administered on the material following the completion of each section of aerodynamics.

Page numbers in this student guide consist of three parts: the unit number (1 for Aerodynamics), followed by a dash (-), the lesson topic number (1 through 12), followed by a dash (-), and the instruction sheet (outline, information or assignment). Page numbers within instruction sheets are denoted at the upper right of the page.

TERMINAL OBJECTIVE

Upon completion of this unit of instruction, the student aviator will demonstrate knowledge of basic aerodynamic factors that affect airplane performance.

Basic Properties of Physics

INTRODUCTION

The purpose of this lesson is to aid the student in understanding basic physics as it relates to aerodynamics.

TERMINAL OBJECTIVE

Upon completion of this unit of instruction, the student aviator will demonstrate knowledge of basic aerodynamic factors that affect airplane performance.

ENABLING OBJECTIVES

1.1   Define scalar quantity, vector, force, mass, volume, density, weight, moment, work, power, energy, potential energy, and kinetic energy.

1.2   State Newton's three Laws of Motion

1.3   Identify examples of Newton's three Laws of Motion.

1.4   Define, compare, and contrast equilibrium and trimmed flight.

1.5   Define static pressure, air density, temperature, lapse rate, humidity, viscosity, and local speed of sound.

1.6   State the relationship between humidity and air density.

1.7   State the relationship between temperature and viscosity.

1.8   State the relationship between temperature and local speed of sound.

1.9   State the pressure, temperature, lapse rate, and air density at sea level in the standard atmosphere using both Metric and English units of measurement.

1.10   State the relationships between altitude and temperature, pressure, air density, and local speed of sound within the standard atmosphere.

1.11   State the relationships between pressure, temperature, and air density using the General Gas Law.

Basic Properties of Physics

INTRODUCTION

This lesson topic will introduce the basic physical laws that govern how an airplane flies.

REFERENCES

1. Aerodynamics for Naval Aviators

2. Aerodynamics for Pilots

3. Introduction to the Aerodynamics of Flight

4. U.S. Standard Atmosphere, 1976

INFORMATION

MATHEMATICAL SYSTEMS

A **scalar** is a quantity that represents only magnitude, e.g., time, temperature, or volume. It is expressed using a single number, including any units. A **vector** is a quantity that represents magnitude and direction. It is commonly used to represent displacement, velocity, acceleration, or force. **Displacement (s)** is the distance and direction of a body's movement (an airplane flies east 100 nm). **Velocity (V)** is the speed and direction of a body's motion, the rate of change of position (an airplane flies south at 400 knots). **Speed** is a scalar equal to the magnitude of the velocity vector. **Acceleration (a)** is the rate and direction of a body's change of velocity (gravity accelerates bodies toward the center of the earth at 32.174 ft/s^2). A **force (F)** is a push or pull exerted on a body (1,000 lbs of thrust pushes a jet through the sky).

A vector may be represented graphically by an arrow. The length of the arrow represents the magnitude and the heading of the arrow represents the direction. Vectors may be added by placing the head of the first vector on the tail of the second and drawing a third vector from the tail of the first to the head of the second. This new vector (Figure 1-1-1) is the resulting magnitude and direction of the original two vectors working together.

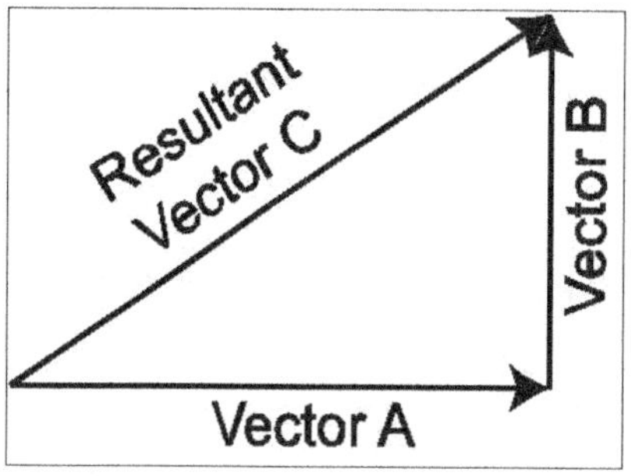

Figure 1-1-1 Vector Addition

DEFINITIONS

Mass (m) is the quantity of molecular material that comprises an object.

Volume (v) is the amount of space occupied by an object.

Density (ρ) is mass per unit volume. It is expressed:

$$\rho = \frac{mass}{volume}$$

Weight (W) is the force with which a mass is attracted toward the center of the earth by gravity.

Force (F) is mass times acceleration:

$$F = m \cdot a$$

A **moment (M)** is created when a force is applied at some distance from an axis or fulcrum, and tends to produce rotation about that point. A moment is a vector quantity equal to a force (F) times the distance (d) from the point of rotation that is perpendicular to the force (Figure 1-1-2). This perpendicular distance is called the moment arm.

Work (W) is done when a force acts on a body and moves it. It is a scalar quantity equal to the force (F) times the distance of displacement (s).

$$W = F \cdot s$$

Power (P) is the rate of doing work or work done per unit of time.

$$P = \frac{W}{t}$$

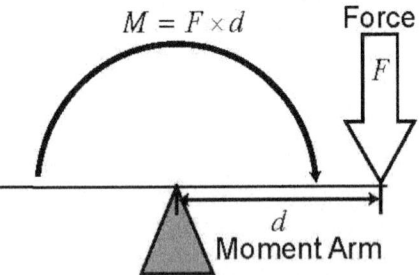

Figure 1-1-2 Moment

Energy is a scalar measure of a body's capacity to do work. There are two types of energy: potential energy and kinetic energy. Energy cannot be created or destroyed, but may be transformed from one form to another. This principle is called conservation of energy. The equation for total energy is:

$$TE = PE + KE$$

Potential energy (PE) is the ability of a body to do work because of its position or state of being. It is a function of mass (m), gravity (g), and height (h):

$$PE = weight \cdot height = mgh$$

Kinetic energy (KE) is the ability of a body to do work because of its motion. It is a function of mass (m) and velocity (V):

$$KE = \tfrac{1}{2}mV^2$$

Work may be performed on a body to change its position and give it potential energy or work may give the body motion so that it has kinetic energy. Under ideal conditions, potential energy may be completely converted to kinetic energy, and vice versa. The kinetic energy of a glider in forward flight is converted into potential energy in a climb. As the glider's velocity (KE) diminishes, its altitude (PE) increases.

NEWTON'S LAWS OF MOTION

NEWTON'S FIRST LAW - THE LAW OF EQUILIBRIUM

"A body at rest tends to remain at rest and a body in motion tends to remain in motion in a straight line at a constant velocity unless acted upon by some unbalanced force."

The tendency of a body to remain in its condition of rest or motion is called inertia. **Equilibrium** is the absence of acceleration, either linear or angular. **Equilibrium flight** exists when the sum of all forces <u>and</u> the sum of all moments around the center of gravity are equal to zero. An airplane in straight and level flight at a constant velocity is acted upon by four forces: thrust, drag, lift and weight. When these forces exactly cancel each other out, the airplane is in equilibrium (Figure 1-1-3).

Trimmed flight exists when the sum of all moments around the center of gravity is equal to zero. In trimmed flight, the sum of the forces may not be equal to zero. For example, an airplane in a constant rate, constant angle of bank turn is in trimmed, but not equilibrium, flight. An airplane in equilibrium flight, however, is always in trimmed flight.

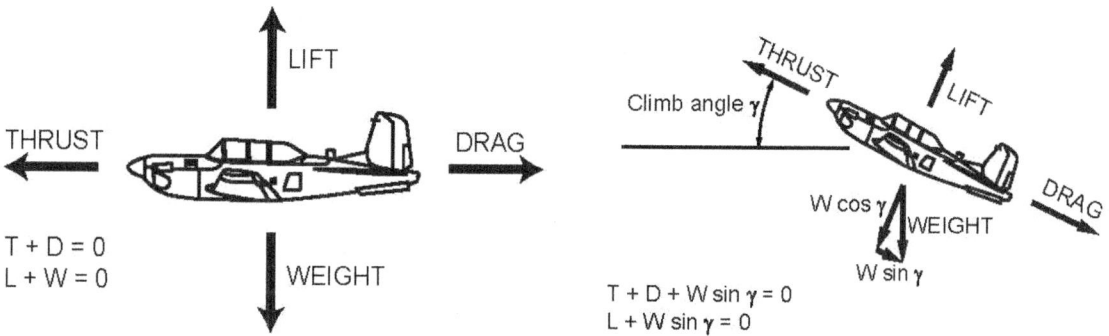

Figure 1-1-3 Equilibrium Level Flight   Figure 1-1-4 Equilibrium Climbing Flight

An airplane does not have to be in straight and level flight to be in equilibrium. Figure 1-1-4 shows an airplane that is climbing, but not accelerating or decelerating, i.e., there are no unbalanced forces. It is another example of equilibrium flight. Thrust must overcome drag plus the parallel component of weight. Lift must overcome the perpendicular component of weight.

An airplane with sufficient thrust to climb vertically at a constant true airspeed can achieve an equilibrium vertical flight condition. Thrust must equal weight plus total drag, and lift must be zero (Figure 1-1-5).

NEWTON'S SECOND LAW - THE LAW OF ACCELERATION

> "An unbalanced force (F) acting on a body produces an acceleration (a) in the direction of the force that is directly proportional to the force and inversely proportional to the mass (m) of the body."

In equation form:

$$a = \frac{F}{m} \qquad a = \frac{V_{out} - V_{in}}{time}$$

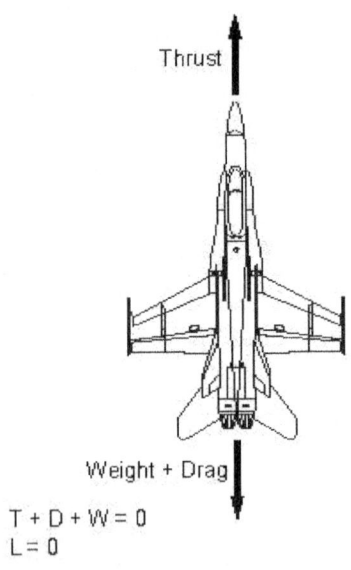

Figure 1-1-5 Equilibrium Vertical Flight

When an airplane's thrust is greater than its drag (in level flight), the excess thrust will accelerate the airplane until drag increases to equal thrust.

NEWTON'S THIRD LAW - THE LAW OF INTERACTION

> "For every action, there is an equal and opposite reaction."

This law is demonstrated by the thrust produced in a jet engine. The hot gases exhausted rearward produce a thrust force acting forward (Figure 1-1-6).

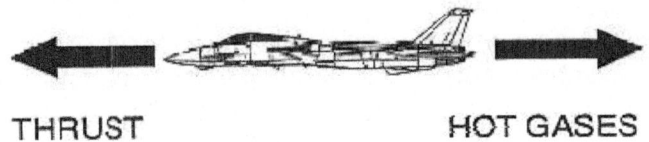

THRUST   HOT GASES

Figure 1-1-6 Action and Reaction

PROPERTIES OF THE ATMOSPHERE

The atmosphere is composed of approximately 78% nitrogen, 21% oxygen, and 1% other gases, including argon and carbon dioxide. Air is considered to be a uniform mixture of these gases, so we will examine its characteristics as a whole rather than as separate gases.

Static pressure (P_S) is the pressure particles of air exert on adjacent bodies. Ambient static pressure is equal to the weight of a column of air over a given area. The force of static pressure always acts perpendicular to any surface that the air particles collide with, regardless of whether the air is moving with respect to that surface.

As altitude increases, there is less air in the column above, so it weighs less. Thus atmospheric static pressure decreases with an increase in altitude. At low altitudes, it decreases at a rate of approximately 1.0 inHg per 1000 ft.

Air density (ρ) is the total mass of air particles per unit of volume. The distance between individual air particles increases with altitude resulting in fewer particles per unit volume. Therefore, air density decreases with an increase in altitude.

Air consists of very many individual particles, each moving randomly with respect to the others. **Temperature (T)** is a measure of the average random kinetic energy of air particles. Air temperature decreases linearly with an increase in altitude at a rate of 2 °C (3.57 °F) per 1000 ft until approximately 36,000 feet. This rate of temperature change is called the **average lapse rate**. From 36,000 feet through approximately 66,000 feet, the air remains at a constant −56.5 °C (−69.7 °F). This layer of constant temperature is called the **isothermal layer**.

Humidity is the amount of water vapor in the air. As humidity increases, water molecules displace an equal number of air molecules. Since water molecules have less mass and do not change the number of particles per unit volume of air, density decreases. Therefore, as humidity increases, air density decreases.

Viscosity (μ) is a measure of the air's resistance to flow and shearing. Air viscosity can be demonstrated by its tendency to stick to a surface. For liquids, as temperature increases, viscosity decreases. Recall that the oil in a car gets thinner when the engine gets hot. Just the opposite happens with air: Air viscosity increases with an increase in temperature.

Sound is caused by disturbances of the air that causes a sudden compression or vibration. This creates a series of alternating compressions and rarefactions which is transmitted to our ears as sound. The compressions and rarefactions are transmitted from one particle to another, but particles do not flow from one point to another. Sound is wave motion, not particle motion. The **local speed of sound** is the rate at which sound waves travel through a particular air mass. The speed of sound, in air, is dependent only on the temperature of the air. The warmer the air, the more excited the particles are in that air mass. The more excited the molecules are, the more easily adjacent molecules can propagate a sound wave. As the temperature of air increases, the speed of sound increases.

THE STANDARD ATMOSPHERE

The atmospheric layer in which most flying is done is an ever-changing environment. Temperature and pressure vary with altitude, season, location, time, and even sunspot activity. It is impractical to take all of these into consideration when discussing airplane performance. In order to disregard these atmospheric changes, an engineering baseline has been developed called the **standard atmosphere**. It is a set of reference conditions giving representative values of air properties as a function of altitude. A summary may be found in Appendix C. Although it is rare to encounter weather conditions that match the standard atmosphere, it is nonetheless representative of average zero humidity conditions at middle latitudes. Unless otherwise stated, any discussion of atmospheric properties in this course will assume standard atmospheric conditions.

	English	Metric (SI)
Static Pressure P_{S0}	29.92 inHg	1013.25 mbar
Temperature T_0	59 °F	15 °C
Average Lapse Rate	3.57 °F / 1000 ft	2 °C / 1000 ft
ρ_0	.0024 slugs / ft^3	1.225 g / l
Local Speed of Sound	661.7 knots	340.4 m / s

Table 1-1-1 Sea Level Standard Atmospheric Conditions

THE GENERAL GAS LAW

The General Gas Law sets the relationship between three properties of air: pressure (P), density (ρ), and temperature (T). It is expressed as an equation where R is a constant for any given gas (such as dry air):

$$P = \rho RT$$

One method to increase pressure is to keep density constant and increase temperature (as in a pressure cooker). If pressure remains constant, there is an inverse relationship between density and temperature. An increase in temperature must result in a decrease in density, and vice versa.

ALTITUDE MEASUREMENT

Altitude is defined as the geometric height above a given plane of reference. **True altitude** is the actual height above mean sea level. **Pressure altitude (PA)** is the height above the standard datum plane. The standard datum plane is the actual elevation at which the barometric pressure is 29.92 inHg. Since the standard datum plane is at sea level in the standard atmosphere, true altitude will be equal to pressure altitude.

Density altitude (DA) is the altitude in the standard atmosphere where the air density is equal to local air density. It is found by correcting pressure altitude for temperature and humidity deviations from the standard atmosphere. In the standard atmosphere, density altitude is equal to pressure altitude. But as temperature or humidity increase, the air becomes less dense, with the effect that the actual air density at one altitude is equal to that of a higher altitude on a standard day. A high DA indicates a low air density.

Density altitude is not used as a height reference, but as a predictor of aircraft performance. A high DA will decrease the power produced by an engine because less oxygen is available for combustion. It will also reduce the thrust produced by a propeller or jet engine because fewer air molecules are available to be accelerated. The reduced power and thrust will reduce an airplane's acceleration and climb performance. A high DA also requires a higher true airspeed for takeoff and landing and will therefore increase takeoff and landing distances.

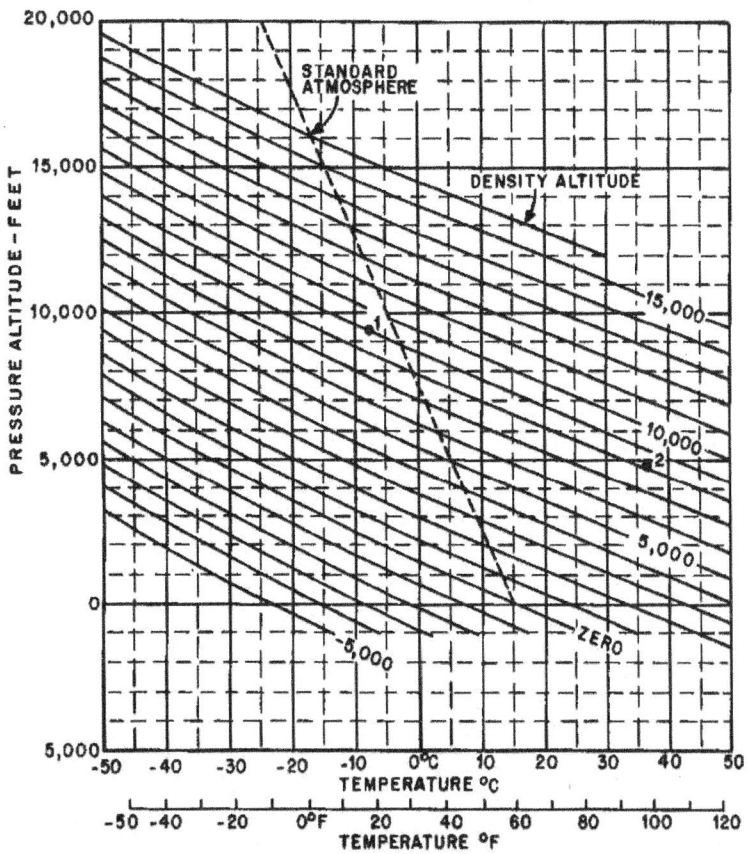

Figure 1-1-7 Density Altitude as a Function of Temperature and Pressure Altitude

Over a typical day, static pressure and pressure altitude remain virtually constant. However, as the sun heats the air, the reduced density causes a dramatic increase in density altitude. This will have a noticeable impact on aircraft performance. Figure 1-1-7 can be used to determine density altitude from pressure altitude and temperature (but does not take into account the effects of humidity).

1.   How does a vector quantity differ from a scalar quantity?

2.   Define mass.

3.   Define weight.

4.   Define air density.

5.   How are a force and a moment related?

6.   Define work. How is it calculated?

7.   Define power.

8.   Define energy. What is the equation for total energy?

9.   Define potential energy (PE).

10.   Define kinetic energy (KE).

11.   State Newton's First Law of Motion.

12.   Under what conditions can both an airplane traveling at a constant speed and direction and an airplane parked on the flight line be in equilibrium?

13.   What is the difference between trimmed flight and equilibrium flight?

14.   State Newton's Second Law of Motion.

15.   State Newton's Third Law of Motion.

16.   Define static pressure. What change in atmospheric static pressure (P_s) occurs with an increase in altitude?

17.   What change in air density occurs with an increase in altitude?

18.   Define air temperature.

19.   What change in air temperature occurs in the standard atmosphere from sea level through 66,000 feet?

20.   What change in air density occurs with an increase in humidity?

21.   Define air viscosity. What change in air viscosity occurs with an increase in temperature?

22.   What is the primary factor affecting the speed of sound in air?

23.   What are the sea level conditions in the standard atmosphere?

24.   State the General Gas Law. What is the relationship between temperature, pressure, and density according to the General Gas Law?

Aircraft Terminology

INTRODUCTION

The purposes of this lesson are to introduce the student to basic aircraft terminology and to describe the physical characteristics of the T-34C.

TERMINAL OBJECTIVE

Upon completion of this unit of instruction, the student aviator will demonstrate knowledge of basic aerodynamic factors that affect airplane performance.

ENABLING OBJECTIVES

1.12	Define, compare, and contrast an aircraft and an airplane.
1.13	List and describe the three major control surfaces of an airplane.
1.14	List and define the five major components of an airplane.
1.15	List and define the components of the airplane reference system.
1.16	Describe the orientation between the components of the airplane reference system.
1.17	List and define the motions that occur around the airplane center of gravity.
1.18	Define wingspan, chordline, chord, tip chord, root chord, average chord, wing area, taper, taper ratio, sweep angle, aspect ratio, wing loading, angle of incidence, and dihedral angle.
1.19	Describe and state the advantages of semi-monocoque fuselage construction.
1.20	Describe full cantilever wing construction.

Aircraft Terminology

INTRODUCTION

This lesson defines basic terms used to describe major components of conventional fixed-wing aircraft.

REFERENCES

1. Aerodynamics for Naval Aviators

2. Introduction to the Aerodynamics of Flight

3. T-34C NATOPS Flight Manual

INFORMATION

MAJOR COMPONENTS OF AN AIRPLANE

An **aircraft** is any device used or intended to be used for flight in the air. It is normally supported either by the buoyancy of the structure (e.g. a balloon or dirigible) or by the dynamic reaction of the air against its surfaces (e.g. an airplane, glider or helicopter).

An **airplane** is a heavier than air fixed wing aircraft that is driven by an engine driven propeller or a gas turbine jet and is supported by the dynamic reaction of airflow over its wings. The T-34C is an unpressurized low winged monoplane with a tricycle landing gear and a tandem cockpit. It will be the primary example of a conventional airplane used throughout this course. The components of a conventional airplane are the fuselage, wings, empennage, landing gear, and engine(s).

The **fuselage** is the basic structure of the airplane to which all other components are attached. It is designed to hold passengers, cargo, etc. Three basic fuselage types are possible: Truss, full monocoque, and semi monocoque. The **truss** type consists of a metal or wooden frame over which a light skin is stretched. It is very strong and easily repaired, but quite heavy. **Full monocoque** is extremely light and strong because it consists of only a skin shell which is highly stressed but almost impossible to repair if damaged. **Semi-monocoque** is a modified version of monocoque having skin, transverse frame members, and stringers, which all share in stress loads and may be readily repaired if damaged. The T-34C uses a semi-monocoque fuselage.

The **wing** is an airfoil attached to the fuselage and is designed to produce lift. It may contain fuel cells, engine nacelles, and landing gear. **Ailerons** are control surfaces attached to the wing to control roll. **Flaps** and slots are high lift devices attached to the wing to increase lift at low airspeeds. The T-34C has a single low-mounted wing with slotted flaps integrated into the trailing edge inboard of the ailerons. Since all bracing is internal, the wings are considered to be **full cantilever**.

The **empennage** is the assembly of stabilizing and control surfaces on the tail of an airplane. It provides the greatest stabilizing influence of all the components of the conventional airplane. The empennage consists of the aft part of the fuselage, the vertical stabilizer, and the horizontal stabilizer. The **rudder** is the upright control surface attached to the vertical stabilizer to control yaw. **Elevators** are the horizontal control surfaces attached to the horizontal stabilizer to control pitch.

The **landing gear** permits ground taxi operation and absorbs the shock encountered during takeoff and landing. The T-34C has tricycle landing gear that includes a nosewheel and two main wheels. During taxi operations, the nosewheel casters; the airplane is steered using its rudder and/or differential braking.

The **engine** provides the thrust necessary for powered flight. Military and commercial airplanes may be fitted with multiple turboprop, turbojet, or turbofan engines. The type of engine depends on the mission requirements of the aircraft. The T-34C has a PT6A-25 turboprop engine.

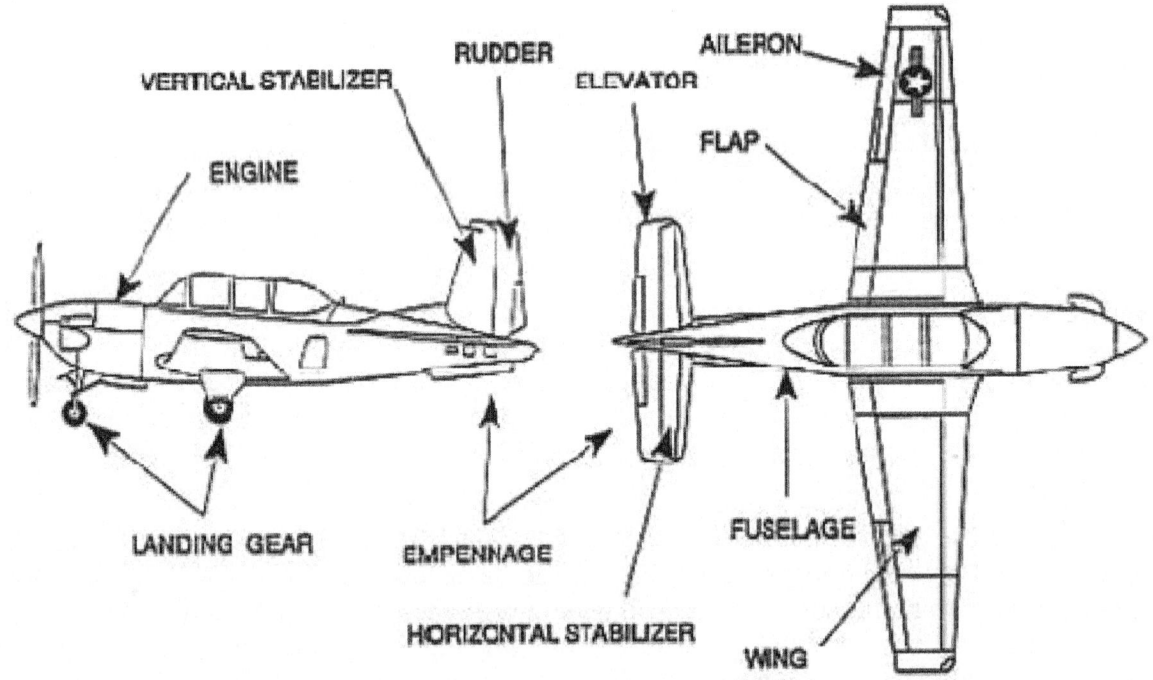

Figure 1-2-1 Airplane Components

AIRPLANE REFERENCE SYSTEM

An airplane's reference system consists of three mutually perpendicular lines (axes) intersecting at a point. This point, called the **center of gravity (CG)** is the point at which all weight is considered to be concentrated and about which all forces and moments (yaw, pitch and roll) are measured. Theoretically, the airplane will balance if suspended at the center of gravity. As fuel burns, ordnance is expended or cargo shifts, the CG will move.

The **longitudinal axis** passes from the nose to the tail of the airplane. Movement of the lateral axis around the longitudinal axis is called **roll**. The **lateral axis** passes from wingtip to wingtip. Movement of the longitudinal axis around the lateral axis is called **pitch**. The **vertical axis** passes vertically through the center of gravity. Movement of the longitudinal axis around the vertical axis is called **yaw**. As an airplane moves through the air, the axis system also moves. Therefore, the movement of the airplane can be described by the movement of its center of gravity.

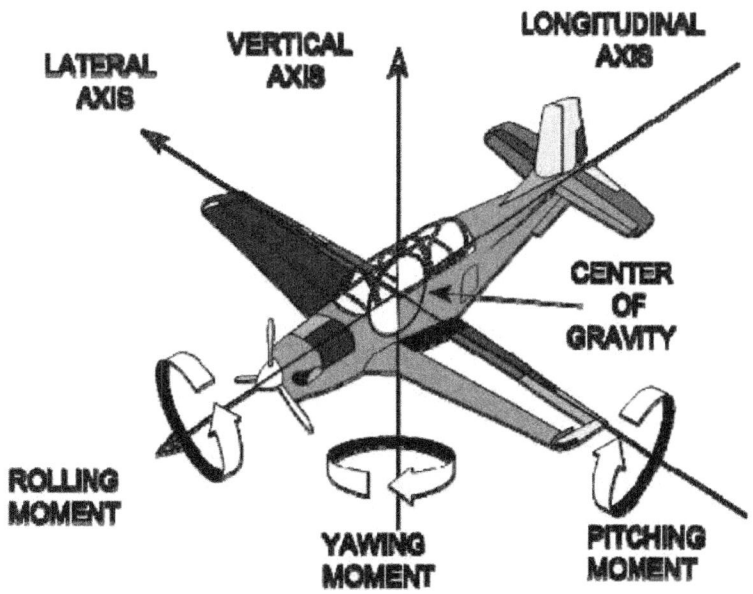

Figure 1-2-2 Airplane Reference System

DIMENSIONS

Wingspan (b) is the length of a wing, measured from wingtip to wingtip. It always refers to the entire wing, not just the wing on one side of the fuselage. The wingspan of the T-34C is 33'5".

The **chordline** of an airfoil is an infinitely long, straight line which passes through its leading and trailing edges. **Chord** is a measure of the width of an airfoil. It is measured along the chordline and is the distance from the leading edge to the trailing edge. Chord will typically vary from the wingtip to the wing root. The **root chord (c_R)** is the chord at the wing centerline and the **tip chord (c_T)** is measured at the wingtip. The **average chord (c)** is the average of every chord from the wing root to the wingtip.

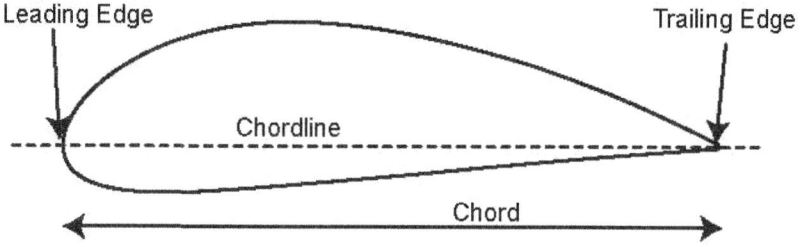

Figure 1-2-3 Wing Cross-Section View

Wing area (S) is the apparent surface area of a wing from wingtip to wingtip. More precisely, it is the area within the outline of a wing in the plane of its chord, including that area within the fuselage, hull or nacelles. The formula for S is:

$$S = bc$$

Taper is the reduction in the chord of an airfoil from root to tip. The wings of the T-34C are tapered to reduce weight, improve structural stiffness, and reduce wingtip vortices. Assuming the wing to have straight leading and trailing edges, taper ratio (λ) is the ratio of the tip chord to the root chord.

$$\lambda = \frac{c_T}{c_R}$$

Sweep angle (Λ) is the angle between the lateral axis and a line drawn 25% aft of the leading edge.

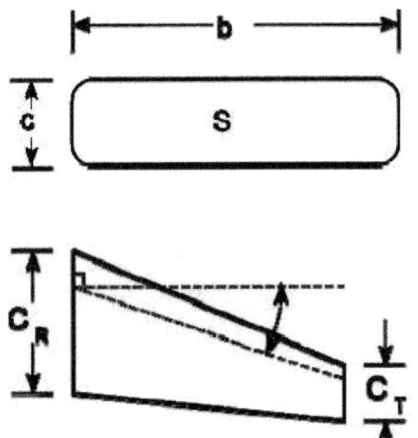

Figure 1-2-4 Wing Planform Views

Aspect ratio (AR) is the ratio of the wingspan to the average chord. An aircraft with a high aspect ratio (35:1), such as a glider, would have a long, slender wing. A low aspect ratio (3:1) indicates a short, stubby wing, such as on a high performance jet.

$$AR = \frac{b}{c}$$

Wing loading (WL) is the ratio of an airplane's weight to the surface area of its wings. There tends to be an inverse relationship between aspect ratio and wing loading. Gliders have high aspect ratios and low wing loading. Fighters with low aspect ratios maneuver at high g-loads and are designed with high wing loading. The wing loading formula is:

$$WL = \frac{W}{S}$$

The **angle of incidence** of a wing is the angle between the airplane's longitudinal axis and the chordline of the wing.

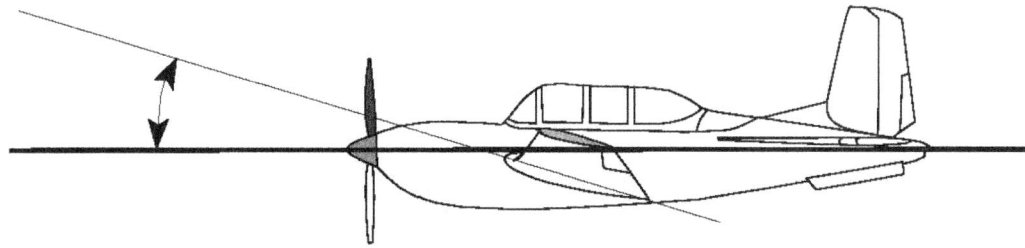

Figure 1-2-5 Angle of Incidence

Dihedral angle is the angle between the spanwise inclination of the wing and the lateral axis. More simply, it is the upward slope of the wing when viewed from the front. A negative dihedral angle is called an **anhedral** angle (sometimes cathedral). The T-34C has dihedral wings to improve lateral stability.

Figure 1-2-6 Dihedral Angle

1.   Define airplane.

2.   What type of construction is used in the fuselage of the T-34C? Why?

3.   What variety of wing has no external bracing?

4.   What control surfaces are attached to the wing?

5.   What control surfaces are attached to the empennage?

6.   What control surface is used for longitudinal control?

7.   What is the primary source of directional control?

8.   Define airplane center of gravity.

9.   List the three airplane axes and the motions that occur about each.

10.   Define wingspan.

11.   What is the difference between chordline, chord, tip chord, root chord and average chord?

12. Define wing area, and state the formula for calculating it.

13. Define taper, taper ratio, and sweep angle.

14. What is aspect ratio? What type of aspect ratio would you expect to find on a B-52 bomber? A high performance fighter?

15. Define angle of incidence. Can the angle of incidence ordinarily be changed?

16. Define wing loading and state the formula for calculating it.

17. Define dihedral angle.

Basic Aerodynamic Principles

INTRODUCTION

The purpose of this lesson is to aid the student in understanding the basic principles of airflow as they relate to aerodynamics.

TERMINAL OBJECTIVE

Upon completion of this unit of instruction, the student aviator will demonstrate knowledge of basic aerodynamic factors that affect airplane performance.

ENABLING OBJECTIVES

1.21	Define steady airflow, streamline, and streamtube.
1.22	Describe the relationship between airflow velocity and cross-sectional area within a streamtube using the continuity equation.
1.23	Describe the relationship between total pressure, static pressure, and dynamic pressure within a streamtube using Bernoulli's equation.
1.24	List the components of the pitot static system.
1.25	State the type of pressure sensed by each component of the pitot static system.
1.26	Define indicated airspeed, calibrated airspeed, equivalent airspeed, true airspeed, and ground speed.
1.27	State the corrections between indicated airspeed, calibrated airspeed, equivalent airspeed, true airspeed, and ground speed.
1.28	Describe the relationships between indicated airspeed, true airspeed, ground speed, and altitude.
1.29	Describe the effects of wind on indicated airspeed, true airspeed, and ground speed.
1.30	Given true airspeed, winds, and time, determine ground speed and distance traveled.
1.31	Define Mach number and critical Mach number.
1.32	Describe the effect of altitude on Mach number and critical Mach number.

Basic Aerodynamic Principles

INTRODUCTION

Before a discussion of the forces of lift and drag, it is important to have an understanding of how air particles and groups of air particles behave.

REFERENCES

1. Aerodynamics for Naval Aviators

2. Aerodynamics for Pilots

3. Introduction to the Aerodynamics of Flight

INFORMATION

PROPERTIES OF AIRFLOW

The atmosphere is a uniform mixture of gases with the properties of a fluid and subject to the laws of fluid motion. Fluids can flow and may be of a liquid or gaseous state. They yield easily to changes in static pressure, density, temperature and velocity. **Steady airflow** exists if at every point in the airflow these four properties remain constant over time. The speed and/or direction of the individual air particles may vary from one

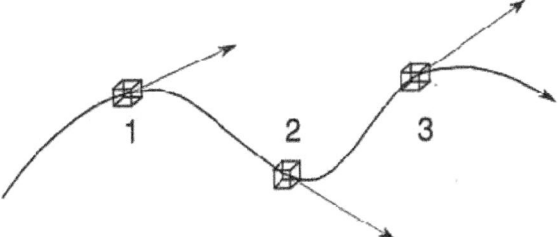

Figure 1-3-1 Streamline in Steady Airflow

point to another in the flow, but the velocity of every particle that passes any given point is always the same. In steady airflow, a particle of air follows the same path as the preceding particle. A **streamline** is the path that air particles follow in steady airflow. In steady airflow, particles do not cross streamlines.

A collection of many adjacent streamlines forms a **streamtube**, which contains a flow just as effectively as a tube with solid walls. In steady airflow, a streamtube is a closed system, in which mass and total energy must remain constant. If mass is added to the streamtube, an equal amount of mass will be removed. An analogy is a garden hose in which each unit of water that flows in displaces another that flows out. Energy cannot be added to or removed from the system, although it can be transformed from one form to another.

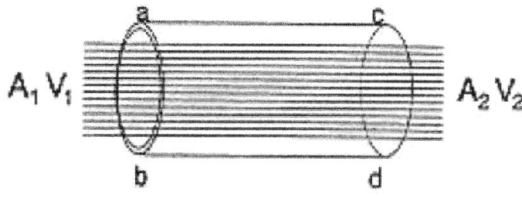

Figure 1-3-2 Streamtube

THE CONTINUITY EQUATION

Let us intersect the streamtube with two planes perpendicular to the airflow at points a-b and c-d, with cross-sectional areas of A_1 and A_2, respectively (Figure 1-3-3). The amount of mass passing any point in the streamtube may be found by multiplying area by velocity to give volume/unit time and then multiplying by density to give mass/unit time. This is called mass flow and is expressed as:

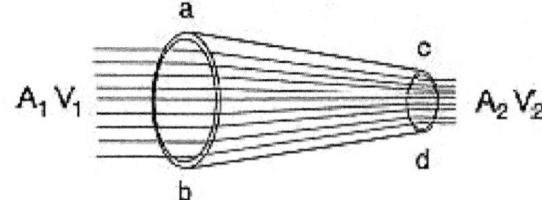

Figure 1-3-3 Continuity of Flow

$$\rho A V$$

The amount of mass flowing through A_1 must equal that flowing through A_2, since no mass can flow through the walls of the streamtube. Thus, an equation expressing the continuity of flow through a streamtube is:

$$\rho_1 A_1 V_1 = \rho_2 A_2 V_2$$

Our discussion is limited to subsonic airflow, so we can ignore changes in density due to compressibility. If we assume that both ends of the streamtube are at the same altitude, then ρ_1 is equal to ρ_2 and we can cancel them from our equation. The simplified continuity equation that we will use is:

$$A_1 V_1 = A_2 V_2$$

If the cross sectional area decreases on one side of the equation, the velocity must increase on the same side so both sides remain equal. Thus velocity and area in a streamtube are inversely related.

BERNOULLI'S EQUATION

Aerodynamics is concerned with the forces acting on a body due to airflow. These forces are the result of pressure and friction. The relationship between pressure and velocity is fundamental to understanding how we create the aerodynamic force on a wing. Bernoulli's equation gives the relationship between the pressure and velocity of steady airflow.

Recall that in a closed system, total energy is the sum of potential energy and kinetic energy, and must remain constant.

Compressed air has potential energy because it can do work by exerting a force on a surface. Therefore, static pressure (P_S) is a measure of potential energy per unit volume.

Moving air has kinetic energy since it can do work by exerting a force on a surface due to its momentum. Dividing KE by volume and substituting ρ for mass/volume gives us dynamic pressure.

$$TE = PE + KE$$

$$\frac{TE}{volume} = \frac{PE}{volume} + \frac{KE}{volume}$$

$$\frac{TE}{volume} = P_S + \frac{\frac{1}{2}mV^2}{volume}$$

$$\frac{TE}{volume} = P_S + \frac{1}{2}\rho V^2$$

$$P_T = P_S + \frac{1}{2}\rho V^2$$

$$P_T = P_S + q$$

Table 1-3-1 Conservation of Energy in a Fluid

Dynamic pressure (q) is the pressure of a fluid resulting from its motion:

$$q = \frac{1}{2}\rho V^2$$

Total pressure (P_T) is the sum of static and dynamic pressure. As with total energy, total pressure also remains constant within a closed system (Table 1-3-1). As area in a streamtube decreases, velocity increases, so q must increase (recall that q depends on V^2). From Bernoulli's equation we know that since q increases, P_S must decrease (Figure 1-3-4). In our streamtube, if dynamic pressure increases, static pressure decreases, and vice versa.

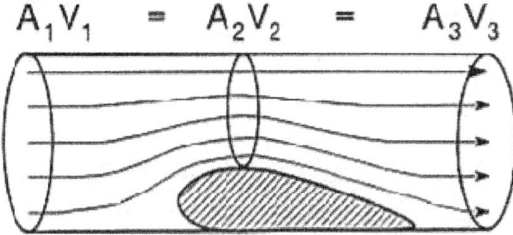

$A_1V_1 \quad = \quad A_2V_2 \quad = \quad A_3V_3$

Figure 1-3-4 Airfoil in a Streamtube

AIRSPEED MEASUREMENT

There are several reasons to measure airspeed. It is necessary to know whether we have sufficient dynamic pressure to create lift, but not enough to cause damage, and velocity is necessary for navigation. If dynamic pressure can be measured, velocity can be calculated. Dynamic pressure cannot be measured directly, but can be derived using Bernoulli's equation as the difference between the total pressure and the static pressure acting on the airplane:

$$q = P_T - P_S$$

The system that accomplishes this is the **pitot static system**. It consists of a pitot tube that senses total pressure (P_T), a static port that senses ambient static pressure (P_S), and a mechanism to compute and display dynamic pressure. We will not concern ourselves with the workings of that mechanism, but simply consider it as a "black box".

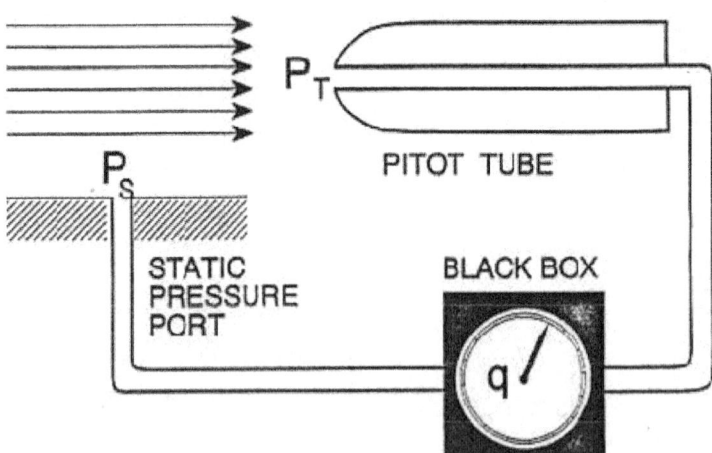

Figure 1-3-5 Pitot Static System

At the entrance to the pitot tube, the airstream has both an ambient static pressure (P_S) and a dynamic pressure (q). Inside the pitot tube, the velocity of the air mass is reduced to zero. As velocity reaches zero, dynamic pressure is converted entirely to static pressure. This converted static pressure is added to the ambient static pressure (P_S) to form a total static pressure equal to the free airstream total pressure (P_T). This total static pressure is connected to one side of a diaphragm inside the black box.

The static pressure port is a hole or series of small holes on the surface of the airplane's fuselage that are flush with the surface. Only ambient static pressure (P_S) affects the static port; no dynamic pressure is sensed. The static port is connected to the other side of the diaphragm in the black box.

The ambient static pressure (P_S) is subtracted from the total pressure (P_T), giving dynamic pressure (q), which is displayed on a pressure gauge inside the cockpit. This gauge is calibrated in **knots of indicated airspeed (KIAS)**. **Indicated airspeed (IAS)** is the instrument indication of the dynamic pressure the airplane is exposed to during flight. To determine true airspeed, certain corrections must be made to IAS.

Instrument error is caused by the static pressure port accumulating erroneous static pressure; slipstream flow causes disturbances at the static pressure port, preventing actual atmospheric pressure measurement. When indicated airspeed is corrected for instrument error, it is called **calibrated airspeed (CAS)**. Often, installation and position error are combined with instrument error. Even the combination of all three errors is usually only a few knots, and is often ignored.

Compressibility error is caused by the ram effect of air in the pitot tube resulting in higher than normal airspeed indications at airspeeds approaching the speed of sound. **Equivalent airspeed (EAS)** is the true airspeed at sea level on a standard day that produces the same dynamic pressure as the actual flight condition. It is found by correcting calibrated airspeed for compressibility error.

True airspeed (TAS) is the actual velocity at which an airplane moves though an air mass. It is found by correcting EAS for density. TAS is EAS corrected for the difference between the local air density (ρ) and the density of the air at sea level on a standard day (ρ_0):

$$\tfrac{1}{2}\rho\left(TAS\right)^2 = \tfrac{1}{2}\rho_0\left(EAS\right)^2$$

$$TAS = \sqrt{\frac{\rho_0}{\rho}}EAS$$

As instrument error is typically small, and compressibility error is minor at subsonic velocities, we will ignore them and develop TAS directly from IAS:

$$TAS = \sqrt{\frac{\rho_0}{\rho}}IAS$$

The pitot static system is calibrated for standard sea level density, so TAS will equal IAS only under standard day, sea level conditions. Since air density decreases with an increase in temperature or altitude, if IAS remains constant while climbing from sea level to some higher altitude, TAS must increase. A rule of thumb is that TAS will be approximately three knots faster than IAS for every thousand feet of altitude.

Ground speed is the airplane's actual speed over the ground. Since TAS is the actual speed of the airplane through the air mass, if we correct TAS for the movement of the air mass (wind), we will have ground speed. It is calculated using the following formulas:

$$GS = TAS - HEADWIND$$

$$GS = TAS + TAILWIND$$

"ICE-TG" is a helpful mnemonic device for the order of the airspeeds.

MACH NUMBER

As an airplane flies, velocity and pressure changes create sound waves in the airflow around the airplane. Since these sound waves travel at the speed of sound, an airplane flying at subsonic airspeeds will travel slower than the sound waves and allow them to dissipate. However, as the airplane nears the speed of sound, these pressure waves "pile up" forming a wall of pressure called a shock wave, which also travels at the speed of sound. As long as the airflow velocity on an airplane remains below the local speed of sound (LSOS), it will not suffer the effects of compressibility. Therefore, it is appropriate to compare the two velocities. **Mach Number (M)** is the ratio of the airplane's true airspeed to the local speed of sound:

$$M = \frac{TAS}{LSOS}$$

Since airplanes accelerate airflow to create lift, there will be local airflow that has a velocity greater than the TAS. Thus an airplane can experience compressibility effects at flight speeds below the speed of sound. **Critical Mach number (M_{CRIT})** is the free airstream Mach number that produces the first evidence of local sonic flow. Simply put, an airplane exceeding M_{CRIT} will have supersonic airflow somewhere on the airplane. Consider a positively cambered airfoil at Mach 0.5. The maximum local airflow velocity on the surface is greater than the true airspeed speed but less than the speed of sound. If an increase to Mach 0.82 boosts the surface airflow velocity up to the local speed of sound, this would be the highest speed possible without supersonic airflow and would determine M_{CRIT}.

	Constant IAS $\overline{IAS} = TAS\sqrt{\dfrac{\rho}{\rho_0}}$	Constant TAS $\overline{TAS} = IAS\sqrt{\dfrac{\rho_0}{\rho}}$	Constant Mach No. $\overline{M} = \dfrac{TAS}{LSOS}$
25,000 ft LSOS = 600 kts	IAS = 200 kts TAS = 300 kts M = 0.5	IAS = 200 kts TAS = 300 kts M = 0.5	IAS = 400 kts TAS = 600 kts M = 1.0
Sea Level LSOS = 661.7 kts	IAS = 200 kts TAS = 200 kts M = 0.3	IAS = 300 kts TAS = 300 kts M = 0.43	IAS = 661.7 kts TAS = 661.7 kts M = 1.0

Table 1-3-2 Effects of Altitude on IAS, TAS and Mach Number
(some values approximated)

1.   State the continuity equation. What are the variables in the equation? When may the density variable be cancelled?

2.   According to the continuity equation, if the velocity of an incompressible fluid is to double, what must happen to the cross sectional area of the flow?

3.   State Bernoulli's equation. Under what conditions does total pressure remain constant? If P_T is constant, how do q and P_S relate?

4.   Describe how the pitot static system works using Bernoulli's equation.

5.   For a given altitude, what is true about the pressure in the static pressure port of the airspeed indicator?

6.   Define IAS and TAS. What is the equation relating the two?

7.   When will IAS equal TAS? How do IAS and TAS vary with increases in altitude?

8.   What must a pilot do to maintain a constant true airspeed during a climb?

9.   An airplane is flying at a six nautical mile per minute ground speed. If it has a 100 knot tailwind, what is its TAS?

10.   An F/A-18 is flying at an eight nautical mile per minute ground speed. If it has a TAS of 600 knots, does it have a headwind or tailwind and how much of one?

11.   Define Mach number and critical Mach Number (M_{CRIT}).

12.   A T-45 is climbing at a constant 350 KIAS. What would be the effect on Mach Number as it climbs? Why?

Lift and Stalls

INTRODUCTION

The purpose of this lesson is to aid the student in understanding lift and stalls as they relate to aerodynamics.

TERMINAL OBJECTIVE

Upon completion of this unit of instruction, the student aviator will demonstrate knowledge of basic aerodynamic factors that affect airplane performance.

ENABLING OBJECTIVES

1.33   Define pitch attitude, flight path, relative wind, angle of attack, mean camber line, positive camber airfoil, negative camber airfoil, symmetric airfoil, aerodynamic center, airfoil thickness, spanwise flow, chordwise flow, aerodynamic force, lift and drag.

1.34   Describe the effects on dynamic pressure, static pressure, and the aerodynamic force as air flows around a cambered airfoil and a symmetric airfoil.

1.35   Describe the effects of changes in angle of attack on the pressure distribution and aerodynamic force of cambered and symmetric airfoils.

1.36   Describe the effects of changes in density, velocity, surface area, camber, and angle of attack on lift.

1.37   List the factors affecting lift that the pilot can directly control.

1.38   Compare and contrast the coefficients of lift generated by cambered and symmetric airfoils.

1.39   Describe the relationships between weight, lift, velocity, and angle of attack in order to maintain straight and level flight, using the lift equation.

1.40   Define boundary layer.

1.41   List and describe the types of boundary layer airflow.

1.42   State the advantages and disadvantages of each type of boundary layer airflow.

1.43   State the cause and effects of boundary layer separation.

1.44   Define stall and state the cause of a stall.

1.45   Define and state the importance of C_{Lmax} and C_{Lmax} AOA.

1.46   State the procedures for stall recovery.

1.47   List common methods of stall warning, and identify those used on the T-34C.

1.48   State the stalling angle of attack of the T-34C.

1.49   Define stall speed.

1.50   Describe the effects of weight, altitude, and thrust on true and indicated stall speed, using the appropriate equation.

1.51   State the purpose of high lift devices.

1.52   State the effect of boundary layer control devices on the coefficient of lift, stalling AOA, and stall speed.

1.53   Describe different types of boundary layer control devices.

1.54   Describe the operation of boundary layer control devices.

1.55   State the effect of flaps on the coefficient of lift, stalling AOA, and stall speed.

1.56   Describe different types of flaps.

1.57   State the methods used by each type of flap to increase the coefficient of lift.

1.58   State the stall pattern exhibited by rectangular, elliptical, moderate taper, high taper, and swept wing planforms.

1.59   State the advantages and disadvantages of tapering the wings of the T-34C.

1.60   State the purpose of wing tailoring.

1.61   Describe different methods of wing tailoring.

1.62   State the types of wing tailoring used on the T-34C.

Lift and Stalls

INTRODUCTION

Lift must overcome the airplane's weight to achieve and maintain equilibrium flight. Understanding how lift is generated and the effect of each of its factors is essential.

REFERENCES

1.   Aerodynamics for Naval Aviators

2.   Aerodynamics for Pilots

3.   Introduction to the Aerodynamics of Flight

4.   T-34C NATOPS Flight Manual

INFORMATION

AIRFOIL TERMINOLOGY

Pitch attitude (θ) is the angle between an airplane's longitudinal axis and the horizon. An airplane's **flight path** is the path described by its center of gravity as it moves through an air mass. **Relative wind** is the airflow the airplane experiences as it moves through the air. It is equal in magnitude and opposite in direction to the flight path.

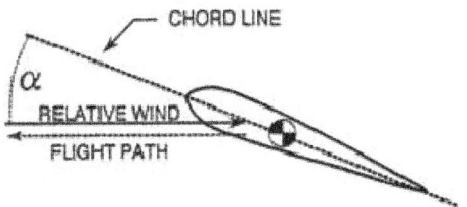

Figure 1-4-1 Flight Path, Relative Wind and Angle of Attack

Angle of attack (α) is the angle between the relative wind and the chordline of an airfoil. Angle of attack is often abbreviated AOA. Flight path, relative wind, and angle of attack should never be inferred from pitch attitude.

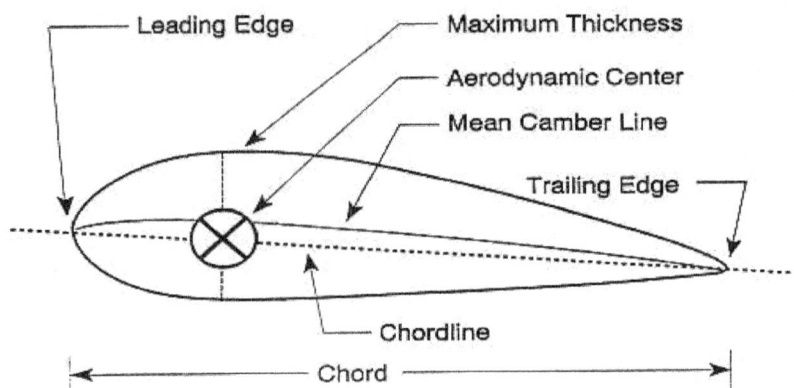

Figure 1-4-2 Airfoil Terminology

The **mean camber line** is a line drawn halfway between the upper and lower surfaces. If the mean camber line is above the chordline, the airfoil has **positive camber**. If it is below the chordline, the airfoil has **negative camber**. If the mean camber line is coincident with the chordline, the airfoil is a **symmetric airfoil**. Airfoil **thickness** is the height of the airfoil profile.

The **aerodynamic center** is the point along the chordline around which all changes in the aerodynamic force take place. On a subsonic airfoil, the aerodynamic center is located approximately one-quarter (between 23% and 27%) of the length of the chord from the leading edge. The aerodynamic center will remain essentially stationary unless the airflow over the wings approaches the speed of sound. Transonic and supersonic flight are not discussed in course.

Spanwise flow is airflow that travels along the span of the wing, parallel to the leading edge. Spanwise flow is normally from the root to the tip. This airflow is not accelerated over the wing and therefore produces no lift.

Chordwise flow is air flowing at right angles to the leading edge of an airfoil. Since chordwise flow is the only flow that accelerates over a wing, it is the only airflow that produces lift.

AERODYNAMIC FORCES

The **aerodynamic force (AF)** is the net force that results from pressure and friction distribution over an airfoil, and comes from two components, lift and drag. **Lift (L)** is the component of the aerodynamic force acting perpendicular to the relative wind. **Drag (D)** is the component of the aerodynamic force acting parallel to and in the same direction as the relative wind.

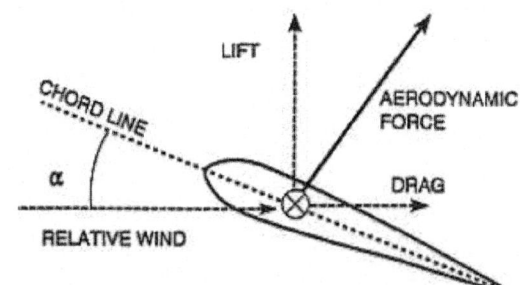

Figure 1-4-3 Aerodynamic Forces

Lift and drag are produced by different physical processes. Lift is produced by a lower pressure distribution on the top of an airfoil than on the bottom.

Drag results from a combination of friction effects and a lower pressure distribution behind an airfoil than in front, and will be discussed in the next lesson. These changes in pressure, along with friction, are responsible for the net aerodynamic force on an airfoil.

Both theoretical and experimental results have shown that both lift and drag can be expressed as the product of dynamic pressure (q), the airfoil surface area (S) and some coefficient that represents the shape and orientation of the airfoil. The **coefficient of lift (C_L)** and the **coefficient of drag (C_D)** are different. The equations for lift and drag are:

$$L = qSC_L = \tfrac{1}{2}\rho V^2 SC_L$$
$$D = qSC_D = \tfrac{1}{2}\rho V^2 SC_D$$

Like its two components, the aerodynamic force can be expressed in the same manner using the **coefficient of force (C_F)**:

$$AF = qSC_F = \tfrac{1}{2}\rho V^2 SC_F$$

LIFT

PRODUCTION OF LIFT

A simplifying assumption made here to ease the discussion of lift is that the air has zero viscosity. Such a gas is referred to as an ideal fluid, and is not subject to friction effects.

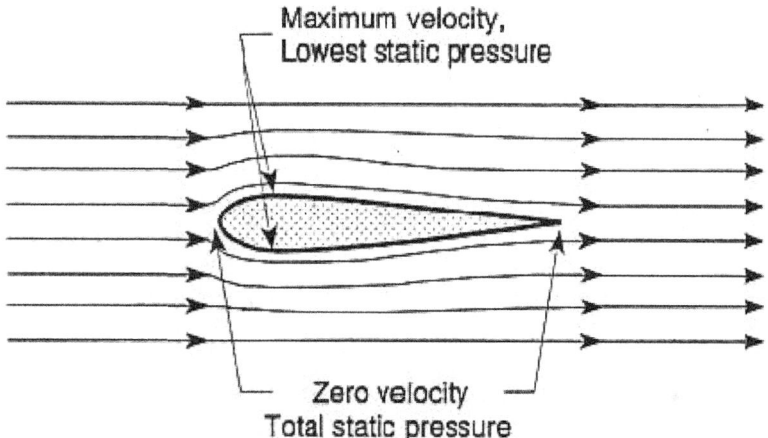

Figure 1-4-4 Airflow Around a Symmetric Airfoil

Airflow around a symmetric airfoil at zero angle of attack will have a streamline pattern similar to that in Figure 1-4-4. As the air strikes the leading edge of the airfoil, its velocity will slow to zero at a point called the leading edge stagnation point. In the area around this point, static pressure is very high. The airflow then separates so that some air moves over the airfoil and some under it, creating two streamtubes. Airflow leaving the area of the leading edge stagnation point will be accelerated due to the decrease in the area of each streamtube. The airflow on both surfaces will reach a maximum velocity at the point of maximum thickness. The airflow then slows until it reaches the trailing edge, where it again slows to zero at a point called the trailing edge stagnation point. Around the trailing edge stagnation point is another area of high static pressure.

In the areas where the airflow velocity is greater than the free airstream velocity, the dynamic pressure is greater and the static pressure is lower. In the areas where the airflow velocity is lower than the free airstream velocity (in particular near the two stagnation points), the dynamic pressure is lower and the static pressure is higher.

A symmetric airfoil at zero angle of attack produces identical velocity increases and static pressure decreases on both the upper and lower surfaces. Since there is no pressure differential perpendicular to the relative wind, the airfoil produces zero net lift. The arrows in Figure 1-4-5 indicate static pressure relative to ambient static pressure. Arrows pointing toward the airfoils indicate higher static pressure; arrows pointing away from the airfoils indicate lower static pressure.

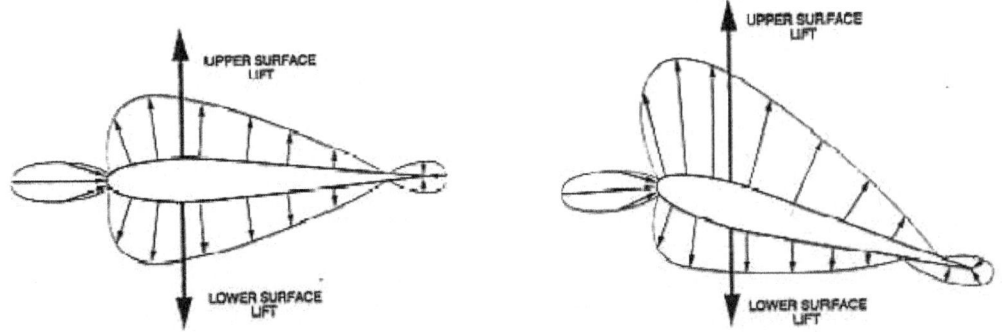

Figure 1-4-5 Pressure Distribution Around Symmetric Airfoil at Zero and Positive AOA

A cambered airfoil is able to produce an uneven pressure distribution even at zero AOA. Because of the positive camber, the area in the streamtube above the wing is smaller than area in the streamtube below the wing and the airflow velocity above the wing is greater than the velocity below the wing.

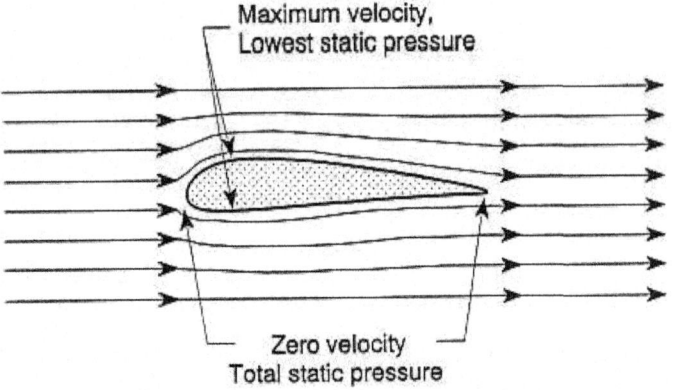

Figure 1-4-6 Airflow Around a Positively Cambered Airfoil

In Figure 1-4-7, the static pressure on both surfaces is less than atmospheric pressure, and thus will produce a lifting force on both upper and lower surfaces. The important point is that these pressures are different. The static pressure on the upper surface will be less than the static pressure on the lower surface, creating a pressure differential. The lower static pressure on the upper surface will "pull" the wing upward, creating a lifting force.

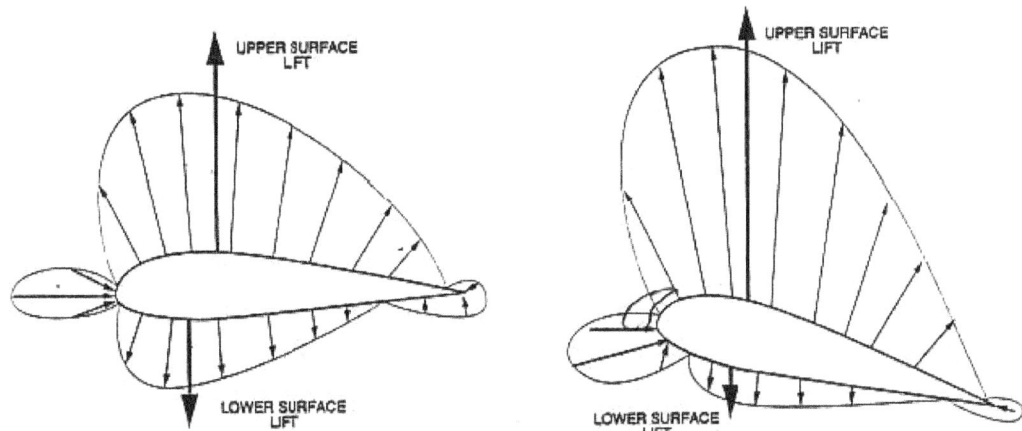

Figure 1-4-7 Pressure Distribution Around Positively Cambered Airfoil at Zero and Positive AOA

As the angle of attack of an airfoil is increased, the leading edge stagnation point will move to a lower point on the leading edge. This shift has the effect of causing the area of the stream-tube above the airfoil to decrease. As with an increase in camber, the velocity of the airflow above the airfoil will increase, lowering the static pressure above the airfoil, increasing the differential pressure and, therefore increasing lift.

FACTORS AFFECTING LIFT

$$L = qSC_L = \tfrac{1}{2}\rho V^2 SC_L$$

There are eight factors that affect lift. The first three are readily apparent: Density (ρ), velocity (V), and surface area (S). The five remaining factors are all accounted for within the coefficient of lift. As stated, both angle of attack (α) and camber affect the production of lift. The remaining three factors are not so easily discernable. They are aspect ratio (AR), viscosity (μ) and compressibility.

When an airfoil is exposed to greater dynamic pressure (q), it encounters more air particles and thus produces more lift. Therefore, lift is dependent upon the density of the air (i.e., the altitude) and the velocity of the airflow. An increase in density or velocity will increase lift.

Since lift is produced by pressure, which is force per unit area, it follows that a greater area produces a greater force. Therefore, an increase in wing surface area produces greater lift.

The pilot has no control over aspect ratio, viscosity and compressibility. Aspect ratio deals with the shape of the wing and will be briefly discussed later. Viscosity affects the aerodynamic force since it decreases the velocity of the airflow immediately adjacent to the wing's surface. Although we consider subsonic airflow to be incompressible, it does compress slightly when it encounters the wing. Because there is no way to control aspect ratio, viscosity, or compressibility, they will be ignored in this discussion unless specifically addressed.

The coefficient of lift depends essentially on the shape of the airfoil and the AOA. Flaps are the devices used to change the camber of an airfoil, and are used primarily for takeoffs and landings. When employed, they will be lowered to a particular setting and remain there until takeoff or landing is complete. This allows us to consider each separate camber situation (i.e.

flap setting) individually and plot C_L against AOA. AOA is the most important factor in the coefficient of lift, and the easiest for the pilot to change.

Figure 1-4-8 plots C_L as it varies with AOA. These curves are for three different airfoils: One symmetric, one negative camber and one positive camber. The shape of the C_L curve is similar for most airfoils. At zero angle of attack, the positive camber airfoil has a positive C_L, and the negative camber airfoil has a negative C_L. The point where the curves cross the horizontal axis is the AOA where the airfoil produces no lift ($C_L = 0$). At zero AOA the symmetric airfoil has $C_L = 0$. The positive camber airfoil must be at a negative AOA, and the negative camber airfoil must be at a positive AOA for the C_L to equal zero.

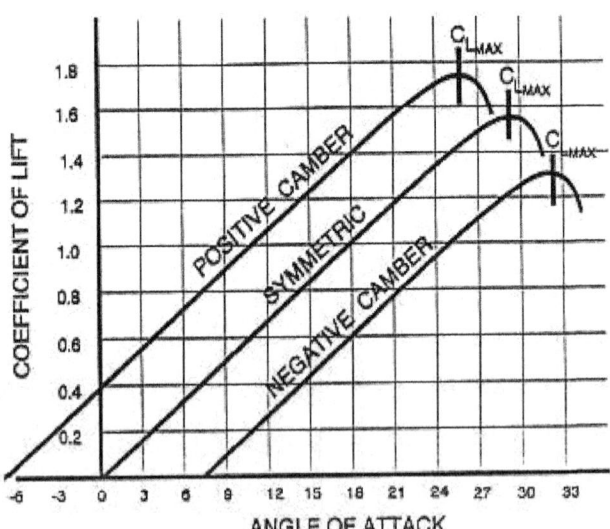

Figure 1-4-8 Camber vs. AOA

As angle of attack increases, the coefficient of lift initially increases. In order to maintain level flight while increasing angle of attack, velocity must decrease. Otherwise, lift will be greater than weight and the airplane will climb. Velocity and angle of attack are inversely related in level flight.

$$L = \tfrac{1}{2} \rho V^2 S C_L$$

As angle of attack continues to increase, the coefficient of lift increases up to a maximum value (C_{Lmax}). The AOA at which C_{Lmax} is reached is called C_{Lmax} AOA. Any increase in angle of attack beyond C_{Lmax} AOA causes a decrease in the coefficient of lift. Since C_{Lmax} is the greatest coefficient of lift that can be produced, we call C_{Lmax} AOA the most *effective* angle of attack. Note that as long as the shape of an airfoil remains constant, C_{Lmax} AOA will remain constant, regardless of weight, dynamic pressure, bank angle, etc.

Although lift is often thought of as an upward force opposing weight, it can act in any direction. It is always perpendicular to the relative wind, not the horizon. In Figure 1-4-9, the relative wind and lift vectors are shown for an airfoil during a loop maneuver. Note that the lift vector is always perpendicular to the relative wind.

STALLS

THE BOUNDARY LAYER

In the preceding discussion of lift, an assumption was made that air was an ideal fluid, with no viscosity or friction effects.

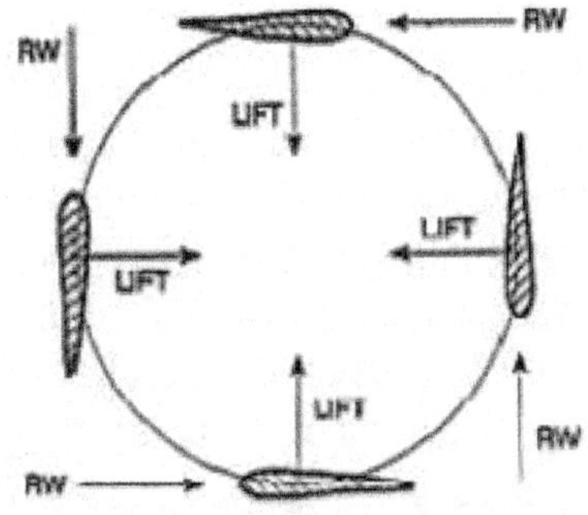

Figure 1-4-9 Lift in a Loop

In actually, when air flows across any surface, friction develops. The air immediately next to the surface slows to near zero velocity as it gives up kinetic energy to friction. As a viscous fluid resists flow or shearing, the adjacent layer of air is also slowed. Succeeding streamlines are slowed less, until eventually some outer streamline reaches the free airstream velocity. The **boundary layer** is that layer of airflow over a surface that demonstrates local airflow retardation due to viscosity. It is usually no more than 1mm thick (the thickness of a playing card) at the leading edge of an airfoil, and grows in thickness as it moves aft over the surface. The boundary layer has two types of airflow.

In **laminar flow**, the air moves smoothly along in streamlines. A laminar boundary layer produces very little friction, but is easily separated from the surface.

In **turbulent flow**, the streamlines break up and the flow is disorganized and irregular. A turbulent boundary layer produces higher friction drag than a laminar boundary layer, but adheres better to the upper surface of the airfoil, delaying boundary layer separation.

Any object that moves through the air will develop a boundary layer that varies in thickness according to the type of surface. The type of flow in the boundary layer depends on its location on the surface. The boundary layer will be laminar only near the leading edge of the airfoil. As the air flows aft, the laminar layer becomes turbulent. The turbulent layer will continue to increase in thickness as it flows aft.

As air flows aft from the leading edge of the airfoil, it moves from a high-pressure area towards the low-pressure area at the point of maximum thickness. This **favorable pressure gradient** assists the boundary layer in adhering to the surface by maintaining its high kinetic energy. As the air flows aft from the point of maximum thickness (lower static pressure) toward the trailing edge (higher static pressure), it encounters an **adverse pressure gradient** which impedes the flow of the boundary layer.

The adverse pressure gradient is strongest at high lift conditions, and at high angles of attack in particular. If the boundary layer does not have sufficient kinetic energy to overcome the adverse pressure gradient, the lower levels of the boundary layer will stagnate. The boundary layer will then separate from the surface, and airflow along the surface aft of the separation point will be reversed. Aft of the separation point, the low static pressure that produced lift is replaced by a turbulent wake.

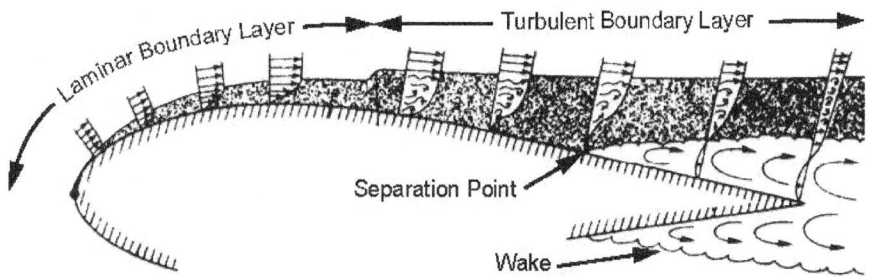

Figure 1-4-10 Boundary Layer Separation

If the separation point moves forward enough close to the leading edge, the net suction on the top of the airfoil will decrease and a decrease in C_L will occur, resulting in a stall. The angle of attack beyond which C_L begins to decrease is C_{Lmax} AOA. Even at low angles of attack there will be a small adverse pressure gradient behind the point of maximum thickness, but it is insignificant compared to the kinetic energy in the boundary layer until C_{Lmax} AOA is approached.

Figure 1-4-11 shows the boundary layer attached at a normal AOA. The point of separation remains essentially stationary near the trailing edge of the wing, until AOA approaches C_{Lmax} AOA. The separation point then progresses forward as AOA increases, eventually causing the airfoil to stall. At high angles of attack the airfoil is similar to a flat plate being forced through the air; the airflow simply cannot conform to the sharp turn. Note that the point where stall occurs is dependent upon AOA and not velocity.

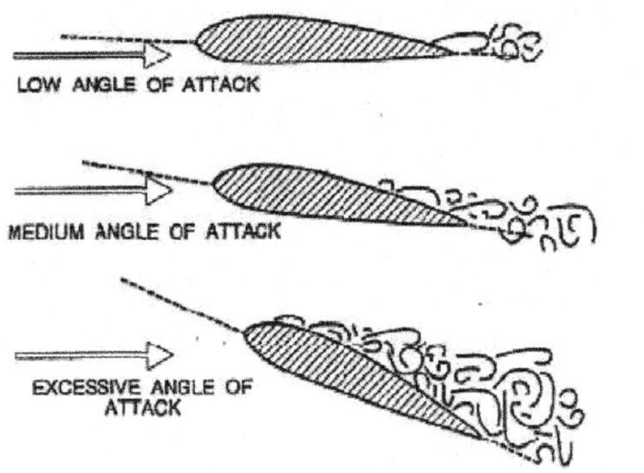

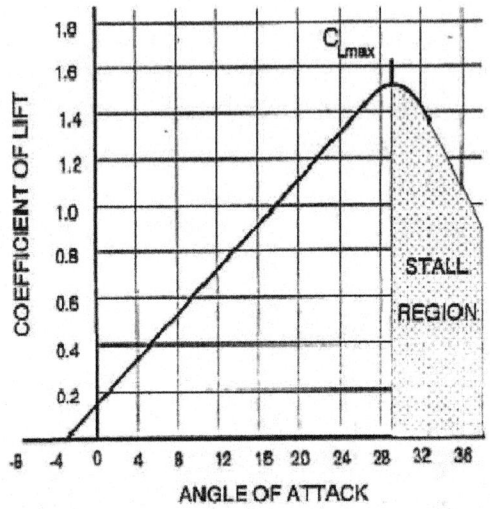

Figure 1-4-11 Progression of Separation Point Forward with Increasing AOA

A **stall** is a condition of flight in which an increase in AOA results in a decrease in C_L. In Figure 1-4-12 C_L increases linearly over a large range of angles of attack then reaches a peak and begins to decrease. The highest value of C_L is referred to as C_{Lmax}, and any increase in AOA beyond C_{Lmax} AOA produces a decrease in C_L. Therefore, C_{Lmax} AOA is known as the **stalling angle of attack** or critical angle of attack, and the region beyond C_{Lmax} AOA is the stall region. Regardless of the flight conditions or airspeed, the wing will always stall at the same AOA, C_{Lmax}. The only cause of a stall is excessive AOA. Stalls result in decreased lift, increased drag, and an altitude loss. They are particularly dangerous at low altitude or when allowed to develop into a spin. The only action necessary for stall recovery is to decrease AOA below C_{Lmax} AOA.

STALL INDICATIONS

Numerous devices may give the pilot a warning of an impending stall. They include AOA indicators, rudder pedal shakers, stick shakers, horns, buzzers, warning lights and other devices. Some of these devices receive their input from attitude gyros, accelerometers, or flight data computers, but most receive input from an AOA probe. The AOA probe is mounted on the fuselage or wing and has a transmitter vane that remains aligned with the relative wind. The vane transmits the angle of attack of the relative wind to a cockpit AOA indicator or is used to activate other stall warning devices. Most US military airplanes have standardized AOA indicators graduated in arbitrary units of angle of attack, or graduated from zero to 100 percent.

Figure 1-4-12 C_L vs. AOA

The T-34C AOA indicator is calibrated so that the airplane stalls between 29.0 and 29.5 units angle of attack regardless of airspeed, nose attitude, weight or altitude. The AOA system in the T-34C is self-adjusting to account for differences in full-flap or no-flap stall angles. The T-34C also uses AOA indexer and rudder shakers that receive their input from an AOA probe on the left wing. The rudder pedal shakers are activated at 26.5 units AOA, when airframe buffet occurs. Stalls at idle in a clean configuration are characterized by a nose down pitch with a slight rolling tendency at near full aft stick. The effect of the landing gear on stalls is negligible, but extending the flaps will aggravate the stall characteristics by increasing the rolling tendency. Increased power will degrade the stall characteristics by increasing nose up stall attitude, increasing buffeting and increasing roll tendency.

STALL SPEED

As angle of attack increases, up to C_{Lmax} AOA, true airspeed decreases in level flight. Since C_L decreases beyond C_{Lmax} AOA, true airspeed cannot be decreased any further. Therefore the minimum airspeed required for level flight occurs at C_{Lmax} AOA. Stall speed (V_S) is the minimum true airspeed required to maintain level flight at C_{Lmax} AOA. Although the stall speed may vary, the stalling AOA remains constant for a given airfoil. Since lift and weight are equal in equilibrium flight, weight (W) can be substituted for lift (L) in the lift equation. By solving for velocity (V), we derive a basic equation for stall speed.

$$V_S = \sqrt{\frac{2W}{\rho S C_{Lmax}}}$$

By substituting the stall speed equation into the true airspeed equation and solving for indicated airspeed, we derive the equation for the indicated stall speed (IAS_S). Weight, altitude, power, maneuvering, and configuration greatly affect an airplane's stall speed. Maneuvering will increase stall speed, but will not be discussed until the lesson that deals with turning flight.

$$IAS_S = \sqrt{\frac{2W}{\rho_0 S C_{Lmax}}}$$

As airplane weight decreases stall speed decreases because the amount of lift required to maintain level flight decreases. When an airplane burns fuel or drops ordnance, stall speeds decrease. Carrier pilots often dump fuel before shipboard landings in order to reduce stall speed and approach speed.

A comparison of two identical airplanes at different altitudes illustrates the effect of altitude on stall speed. The airplane at a higher altitude encounters fewer air molecules. In order to create sufficient dynamic pressure to produce the required lift, it must fly at a higher velocity (TAS). Therefore, an increase in altitude will increase stall speed. Since ρ_0 is constant, indicated stall speed will not change as altitude changes.

The stall speed discussed up to this point assumes that aircraft engines are at idle, and is called power-off stall speed. Power-on stall speed will be less than power-off stall speed because at high pitch attitudes, part of the weight of the airplane is actually being supported by the vertical component of the thrust vector. For propeller driven airplanes the portion of the wing immediately behind the propeller produces more lift because the air is being accelerated by the propeller. Power-on stall speed in the T-34C is approximately 9 knots less than power-off stall speed.

$$V_S = \sqrt{\frac{2(W - T\sin\theta)}{\rho S C_L}} \qquad IAS_S = \sqrt{\frac{2(W - T\sin\theta)}{\rho_0 S C_L}}$$

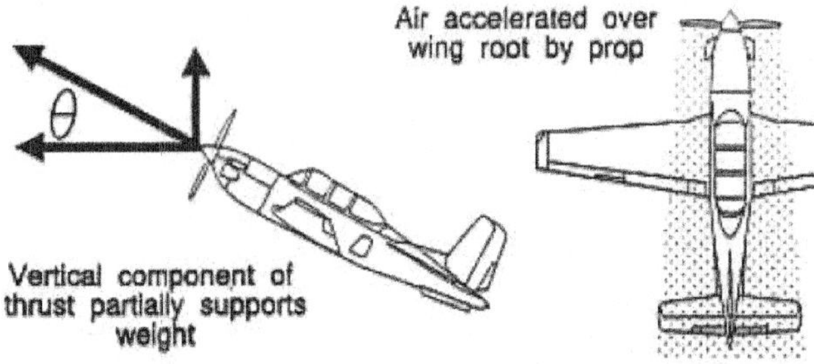

θ
Vertical component of thrust partially supports weight
Air accelerated over wing root by prop

Figure 1-4-13 Power-On Stall

HIGH LIFT DEVICES

High lift devices also affect stall speeds since they increase C_L at high AOA. The primary purpose of high lift devices is to reduce takeoff and landing speeds by reducing stall speed. The increase in C_L allows a decrease in airspeed. For example, an airplane weighing 20,000 pounds flying at 250 knots develops 20,000 pounds of lift. As the airplane slows to 125 knots for landing, high lift devices can increase C_L so that 20,000 pounds of lift can still be produced at the lower velocity. There are two common types of high lift devices: Those that delay boundary layer separation, and those that increase camber.

BOUNDARY LAYER CONTROL DEVICES

The maximum value of C_L is limited by the AOA at which boundary layer separation occurs. If airflow separation can be delayed to an AOA higher than normal stalling AOA, a higher C_{Lmax} can be achieved. Both C_{Lmax} and C_{Lmax} AOA increase with the use of Boundary Layer Control (BLC) devices.

COEFFICIENT OF LIFT
BLC
BASIC AIRFOIL
ANGLE OF ATTACK

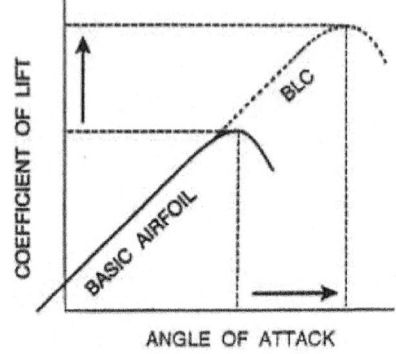

Figure 1-4-14 Effect of BLC

Slots operate by allowing the high static pressure air beneath the wing to be accelerated through a nozzle and injected into the boundary layer on the upper surface of the airfoil. As the air is accelerated through the nozzle, its potential energy is converted to kinetic energy. Using this extra kinetic energy, the turbulent boundary layer is able to overcome the adverse pressure gradient and adhere to the airfoil at higher AOAs. There are generally two types of slots, fixed slots and automatic slots.

SLAT
SLOT

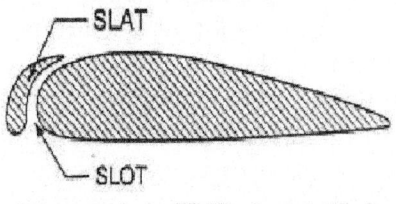

Figure 1-4-15 Slat and Slot

Fixed slots are gaps located at the leading edge of a wing that allow air to flow from below the wing to the upper surface. High pressure air from the vicinity of the leading edge stagnation point is directed through the slot, which acts as a nozzle converting the static pressure into dynamic pressure. The high kinetic energy air leaving the nozzle increases the energy of the boundary layer and delays separation. This is very efficient and causes only a small increase in drag.

Slats are moveable leading edge sections used to form **automatic slots**. When the slat deploys, it opens a slot. Some slats are deployed aerodynamically. At low AOA, the slat is held flush against the leading edge by the high static pressure around the leading edge stagnation point. When the airfoil is at a high AOA, the leading edge stagnation point and associated high pressure area move down away from the leading edge and are replaced by a low (suction) pressure which creates a chordwise force forward and actuates the slat. Other automatic slots are deployed mechanically, hydraulically or electrically.

Since slats and slots on their own effect no change in camber, there is no change to C_L at low AOA. The higher value of C_{Lmax} is achieved at a higher AOA, i.e., the stall is delayed to a higher AOA.

A simple form of BLC is achieved by vortex generators, which are small vanes installed on the upper surface of an airfoil to disturb the laminar boundary layer and induce a turbulent boundary layer. This ensures the area behind the vortex generators benefits from airflow that adheres better to the wing, delaying separation.

CAMBER CHANGE

The most common method of increasing C_{Lmax} is increasing the camber of the airfoil. There are various types of high lift devices that increase the camber of the wing and increase C_{Lmax}. Trailing edge flaps are the most common type of high lift devices, but leading edge flaps are not unusual. The change in C_L and AOA due to flaps is shown in Figure 1-4-16. Note the value of C_L for this airfoil before and after flaps are deployed. Extending the flaps increases the airfoil's positive camber, shifting its zero lift point to the left. Note that the stalling AOA (C_{Lmax} AOA) decreases.

Although stalling AOA decreases, visibility on takeoff and landing improves due to flatter takeoff and landing attitudes made possible by these devices. Since boundary layer control devices increase stalling AOA, many modern designs utilize BLC with camber change devices to maintain low pitch attitudes during approach and landing. Flaps also increase the drag on the airplane, enabling a steeper glide slope and higher power setting during approach without increasing the airspeed. This allows an airplane such as an EA-6B to carry more thrust throughout the landing phase and not significantly increase the approach speed (a higher throttle setting results in less spool-up time in case of a wave-off or go-around). For

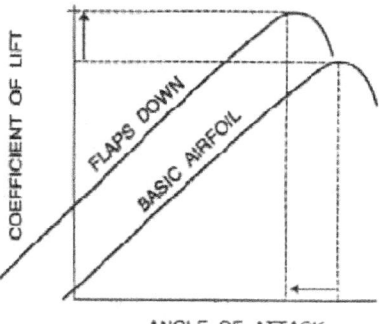

Figure 1-4-16 Effect of Flaps

many airplanes, the first 50 percent of flap down movement produces most of the desired lift increase with less than half of the unwanted drag increase. Thus, raising flaps from 100 to 50 percent reduces drag significantly without a large loss of lift. This is especially important during engine failures on multi-engine airplanes.

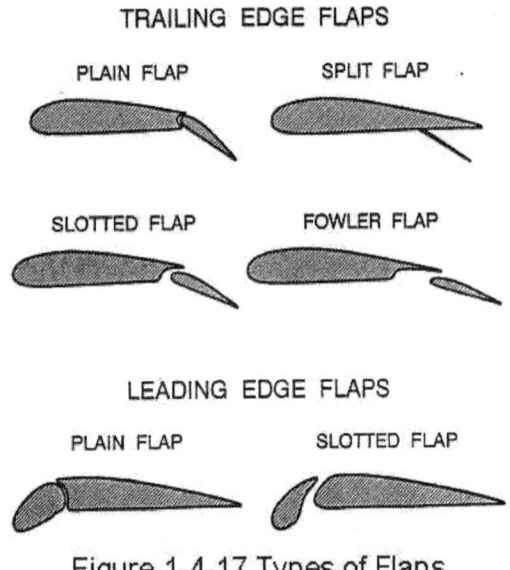

Figure 1-4-17 Types of Flaps

A **plain flap** is a simple hinged portion of the trailing edge that is forced down into the airstream to increase the camber of the airfoil. A **split flap** is a plate deflected from the lower surface of the airfoil. This type of flap creates a lot of drag because of the turbulent air between the wing and deflected surface. A **slotted flap** is similar to the plain flap, but moves away from the wing to open a narrow slot between the flap and wing for boundary layer control. A slotted flap may cause a slight increase in wing area, but the increase is insignificant. The **fowler flap** is used extensively on larger airplanes. When extended, it moves down, increasing the camber, and aft, causing a significant increase in wing area as well as opening one or more slots for boundary layer control. Because of the larger area created on airfoils with fowler flaps, a large twisting moment is developed. This requires a structurally stronger wing to withstand the increased twisting load and precludes their use on high speed, thin wings.

Leading edge flaps are devices that change the wing camber at the leading edge of the airfoil. They may be operated manually with a switch or automatically by computer. Leading edge plain flaps are similar to a trailing edge plain flap. Leading edge slotted flaps are similar to trailing edge slotted flaps, but are sometimes confused with automatic slots. Often the terms are interchangeable since many leading edge devices have some characteristics of both flaps and slats.

The exact stall speed for various airplane conditions are given in stall speed charts in an airplane's flight manual. The directions on how to use the stall speed chart are on the chart itself and are self-explanatory.

STALL PATTERN AND WING DESIGN

The most desirable stall pattern on a wing is one that begins at the root. The primary reason for a root first stall pattern is to maintain aileron effectiveness until the wing is fully stalled. Additionally, turbulent airflow from the wing root may buffet the empennage, providing an aerodynamic warning of impending stall. Different planforms have characteristic stall patterns.

The lift distribution on the **rectangular wing** (λ = 1.0) is due to low lift coefficients at the tip and high lift coefficients at the root. Since the area of the highest lift coefficient will stall first, the rectangular wing has a strong root stall tendency. This pattern provides adequate stall warning and aileron effectiveness. This planform is limited to low speed, light-weight airplanes where simplicity of construction and favorable stall characteristics are the predominating requirements.

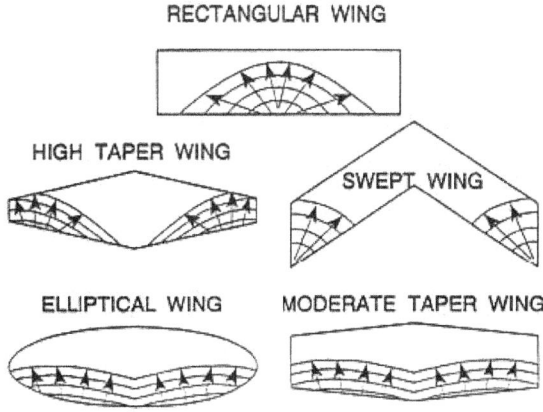

Figure 1-4-18 Stall Patterns

A **highly tapered wing** (λ = 0.25) is desirable from the standpoint of structural weight, stiffness, and wingtip vortices. Tapered wings produce most of the lift toward the tip and have a strong tip stall tendency.

Swept wings are used on high speed aircraft because they reduce drag and allow the airplane to fly at higher Mach numbers. They have a similar lift distribution to a tapered wing, and therefore stall easily and have a strong tip stall tendency. When the wingtip stalls, the stall rapidly progresses over the remainder of the wing.

The **elliptical wing** has an even distribution of lift from the root to the tip and produces minimum induced drag. An even lift distribution means that all sections stall at the same angle of attack. There is little advanced warning and aileron effectiveness may be lost near a stall. It is also more difficult to manufacture than other planforms, but is considered the ideal subsonic wing due to its lift to drag ratio.

Moderate taper wings (λ = 0.5) have a lift distribution and stall pattern that is similar to the elliptical wing. The T-34C uses tapered wings because they reduce weight, improve stiffness, and reduce wingtip vortices. However, the even stall progression is undesirable for the same reasons as with the elliptical wing. As a stall progresses, the pilot will lose lateral control of the airplane.

WING TAILORING

Although stalls cannot be eliminated, they can be made more predictable by having the wing stall gradually. Since most airplanes do not have rectangular wings, they tend to stall with little or no warning. Wing tailoring techniques are used to create a root to tip stall progression and give the pilot some stall warning while ensuring that the ailerons remain effective up to a complete stall. Trailing edge flaps decrease the stalling angles of attack in their vicinity, causing initial stall in the flap area. BLC devices generally delay stall in their vicinity. Propeller-driven airplanes may have a tip stall tendency during power-on stalls due to the increased airflow over the wing root.

Geometric twist is a decrease in angle of incidence from wing root to wingtip. The root section is mounted at some angle to the longitudinal axis, and the leading edge of the remainder of the wing is gradually twisted downward. This results in a decreased AOA at the wingtip due to its lower angle of incidence. The root stalls first because of its higher AOA. The T-34C wing is geometrically twisted 3.1°.

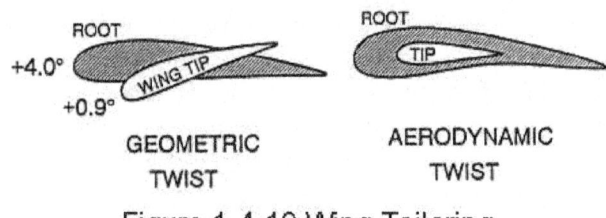

Figure 1-4-19 Wing Tailoring

Aerodynamic twist, also called section variation, is a gradual change in airfoil shape that increases C_{Lmax} AOA to a higher value near the tip than at the root. This can be accomplished by a decrease in camber from the root to the tip and/or by a decrease in the relative thickness of the wing (as compared to chord) from the root to the tip. Since thicker and more positively cambered airfoils stall at lower angles of attack, the wing root stalls before the wingtip. The T-34C wings are aerodynamically twisted.

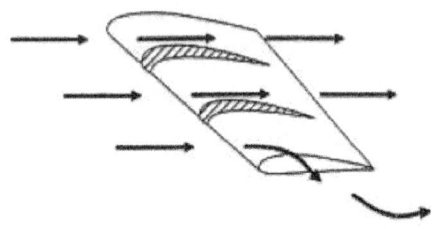

Figure 1-4-20 Stall Fences

The spanwise flow on a swept wing is not accelerated over the wing and does not contribute to the production of lift. Instead, it induces a strong tip stall tendency. **Stall fences** redirect the airflow along the chord, thereby delaying tip stall and enabling the wing to achieve a higher AOA without stalling.

A sharply angled piece of metal called a **stall strip** is mounted on the leading edge of the root section to induce a stall at the wing root. Since subsonic airflow cannot flow easily around sharp corners, it separates the boundary layer at higher angles of attack, ensuring that the root section stalls first. Stall strips are located near the root at the leading edge of the T-34C wing.

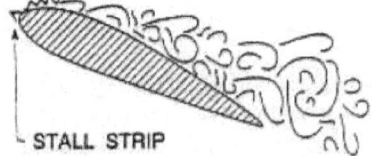

Figure 1-4-21 Stall Strip

STALL RECOVERY

To produce the required lift at slow airspeeds, the pilot must fly at high angles of attack. Because flying slow at high angles of attack is one of the most critical phases of flight, pilots practice recovering from several types of stalls during training.

The steps in a stall recovery involve simultaneously adding power, relaxing back stick pressure and rolling wings level ("Max, relax, level").

The pilot adds power to help increase airspeed, breaking any descent due to the stall (especially at low altitudes) and restoring a velocity greater than V_S.

The pilot must decrease the angle of attack to recover from a stalled condition, as the only reason the aircraft stalled was that it exceeded its stalling angle of attack. The pilot's initial reaction, especially at low altitudes, might be to pull the nose up. However, the exact opposite must be done. The stick must be moved forward to decrease the angle of attack and allow the wing to provide sufficient lift to fly once again. By lowering the nose, angle of attack is de-creased and the boundary layer separation point moves back toward the trailing edge, restoring lift.

The pilot rolls out of bank to wings level to help decrease the stall velocity and use all available lift to break any descent due to the stall.

1.   What is pitch attitude?
      A.   The angle between the chordline and the tip path plane
      B.   The angle between the chordline and the relative wind
      C.   The angle between the longitudinal axis and the relative wind
      D.   The angle between the longitudinal axis and the horizon

2.   Define flight path.

3.   Define relative wind.

4.   How is angle of attack measured?
      A.   Between the top surface of the airfoil and the chordline
      B.   Between the relative wind and the bottom surface of the airfoil
      C.   Between the relative wind and the chordline
      D.   Between the mean camber line and the relative wind

5.   Define mean camber line. How does the mean camber line define the type of airfoil?

6.   Define aerodynamic center.

7.   What is the effect on static pressure of increasing angle of attack on a symmetric airfoil?

8.   Define aerodynamic force, and state the aerodynamic force equation.

9.   What are the two component forces that make up the aerodynamic force?

10.   State the lift equation, and identify the three factors that the pilot can normally control.

11.   In order to maintain level flight while decreasing airspeed, what action must the pilot take?

12.   Draw the C_L curves for both symmetric and a positively cambered airfoils and explain their differences.

13.   At what angle of attack is maximum lift produced?

14.   What is the orientation of the lift vector to the relative wind?

15.   An airplane that has been flying straight and level maintains constant airspeed and increases AOA to a new value below C_{Lmax}. What will happen?

16.   An airplane that has been flying straight and level decreases airspeed and maintains a constant AOA. What will happen?

17.   An airplane that has been flying straight and level increases airspeed and decreases AOA. What will happen?

18.   An airplane that has been flying straight and level increases airspeed and increases AOA to a new value greater than C_{Lmax}. What will happen?

19.   An airplane that has been flying straight and level maintains constant airspeed and decreases AOA. What will happen?

20.   Describe laminar and turbulent flow.

21.   What is the primary feature of airflow separation? During boundary layer separation, how does
        the separation point move along the airfoil?

22.   Define stall. What is the cause of stall?

23.   How does increasing the speed of an airfoil affect its stalling angle of attack?
        A.   Increases the stalling angle of attack
        B.   Reduces the stalling angle of attack
        C.   Has no effect on stalling angle of attack
        D.   Eliminates the stalling angle of attack

24.   In a stall, what is the result of increasing AOA?
        A. C_L increases and lift decreases
        B. C_L increases and lift increases
        C. C_L decreases and lift increases
        D. C_L decreases and lift decreases

25.   Define stall speed and state the normal stall speed equation.

26.   State the relationship of stall speed to gross weight and altitude.

27.   The vertical component of thrust supports a portion of gross weight during a _____ stall,
        reducing stall speed by approximately ____ knots in the T-34C.
        A.   turning, 7
        B.   accelerated, 9
        C.   power on, 9
        D.   normal, 7

28. What is the purpose of high lift devices? How does each type of high lift device affect stalling AOA?

29. Name two boundary layer control devices. In general, how do slots work?

30. What effect does lowering the flaps have on lift and drag?
    A.   Increases lift and increases drag
    B.   Decreases lift and decreases drag
    C.   Increases lift without increasing drag
    D.   Does not affect lift but increases drag

31. List several high lift devices that increase the camber of an airfoil. Which are found on the T-34C? Which produces the greatest increase in C_{Lmax}?

32. Why is it beneficial for the wing root to stall first?

33. Define geometric and aerodynamic twist.

34. What are the general steps in a stall recovery?

Drag

INTRODUCTION

The purpose of this lesson is to aid the student in understanding drag as it relates to aerodynamics.

TERMINAL OBJECTIVE

Upon completion of this unit of instruction, the student aviator will demonstrate knowledge of basic aerodynamic factors that affect airplane performance.

ENABLING OBJECTIVES

1.63   Define total drag, parasite drag, and induced drag.

1.64   List the three major types of parasite drag.

1.65   State the cause of each major type of parasite drag.

1.66   State the aircraft design features that reduce each major type of parasite drag.

1.67   Describe the effects of changes in density, velocity, and equivalent parasite area on parasite drag, using the parasite drag equation.

1.68   Describe the effects of upwash and downwash on the lift generated by an infinite wing.

1.69   Describe the effects of upwash and downwash on the lift generated by a finite wing.

1.70   State the cause of induced drag.

1.71   State the aircraft design features that reduce induced drag.

1.72   Describe the effects of changes in lift, weight, density, and velocity on induced drag, using the induced drag equation.

1.73   Describe the effects of changes in velocity on total drag.

1.74   Define and state the purpose of the lift to drag ratio.

1.75   State the importance of L/D_{MAX}.

Drag

INTRODUCTION

Since thrust must overcome drag for equilibrium flight, drag will play a major role in any discussion of airplane performance. Understanding the causes and effects of drag is essential for the prospective aviator.

REFERENCES

1.   Aerodynamics for Naval Aviators

2.   Aerodynamics for Pilots

3.   Introduction to the Aerodynamics of Flight

INFORMATION

DRAG

Drag is the component of the aerodynamic force that is parallel to the relative wind, and acts in the same direction. The drag equation is the same as the aerodynamic force equation, except that that the coefficient of drag (C_D) is used.

$$D = \tfrac{1}{2}\rho V^2 S C_D$$

C_D may be plotted against angle of attack for a given aircraft with a constant configuration (Figure 1-5-1). Note that C_D is low and nearly constant at very low angles of attack. As angle of attack increases, C_D rapidly increases. Since there is always some resistance to motion, drag will never be zero, so C_D will

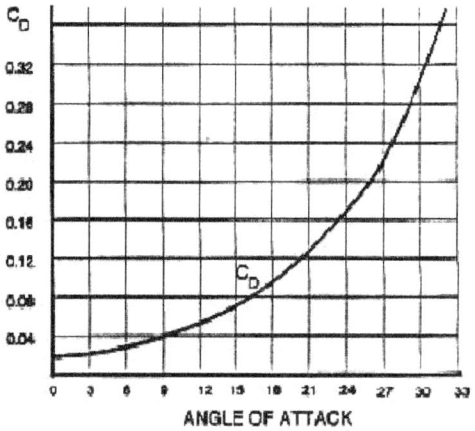

Figure 1-5-1 Coefficient of Drag

never be zero. Drag is divided into parasite drag and induced drag. By independently studying the factors that affect each type, we can better understand how they act when combined.

$$D_T = D_P + D_I$$

PARASITE DRAG

Parasite drag (D_P) is composed of form drag, friction drag and interference drag. It is all drag that is not associated with the production of lift.

Form drag, also known as pressure drag or profile drag, is caused by airflow separation from a surface and the low pressure wake that is created by that separation. It is primarily dependent upon the shape of the object. In Figure 1-5-2, the flat plate has a leading edge stagnation point at the front with a very high static pressure. There is also a low static pressure wake area behind the plate. This pressure differential pulls the plate backward and retards forward motion. Conversely, streamlines flow smoothly over a smooth shape (Figure 1-5-3 and Figure 1-5-4) and less form drag is developed.

To reduce form drag, the fuselage and other surfaces exposed to the airstream are stream-lined (shaped like a teardrop). Streamlining reduces the size of the high static pressure area near the leading edge stagnation point and reduces the size of the low static pressure wake. Because of the decreased pressure differential, form drag is decreased.

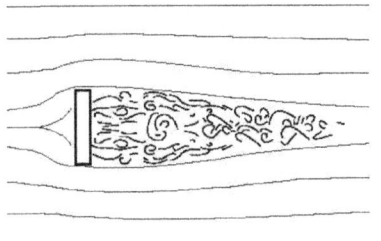
Figure 1-5-2 Flat Plate

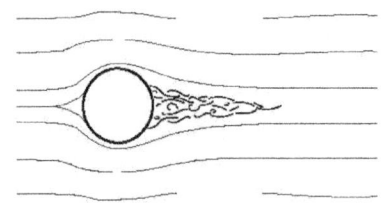
Figure 1-5-3 Sphere

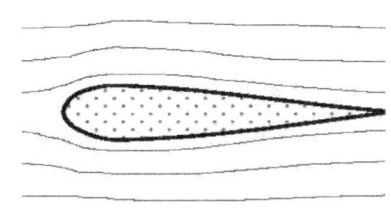
Figure 1-5-4 Streamlining

Due to viscosity, a retarding force called **friction drag** is created in the boundary layer. Turbulent flow creates more friction drag than laminar flow. Friction drag is usually small per unit area, but since the boundary layer covers the entire surface of the airplane, friction drag can become significant in larger airplanes. Rough surfaces increase the thickness of the boundary layer and create greater skin friction.

Friction drag can be reduced by smoothing the exposed surfaces of the airplane through painting, cleaning, waxing or polishing. Since irregularities of the wing's surface cause the boundary layer to become turbulent, using flush rivets on the leading edges also reduces friction.

Since friction drag is much greater in the turbulent boundary layer, it might appear that preventing the laminar flow from becoming turbulent would decrease drag. However, if the boundary layer were all laminar

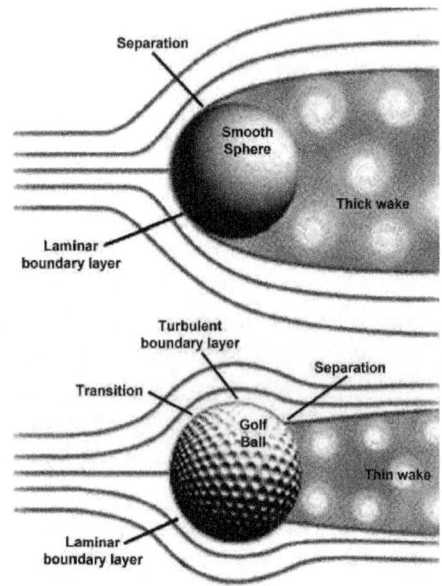
Figure 1-5-5 Dimples on a Golf Ball

airflow, it would easily separate from the surface, creating a large wake behind the airfoil and increasing form drag. Since turbulent airflow adheres to the surface better than laminar flow, maintaining turbulent airflow on an airfoil will significantly reduce form drag with only a small increase in friction. For this reason a golf ball with dimples will go farther than a smooth ball, as it has less form drag.

Interference drag is generated by the mixing of streamlines between components. An example is the air flowing around the fuselage mixing with air flowing around an external fuel tank. We know the drag of the fuselage and the drag of the fuel tank individually. The total drag after we attach the fuel tank will be greater than the sum of the fuselage and the fuel tank separately. Roughly 5 to 10 percent of the total drag on an airplane can be attributed to interference drag. Interference drag can be minimized by proper fairing and filleting, which allows the streamlines to meet gradually rather than abruptly.

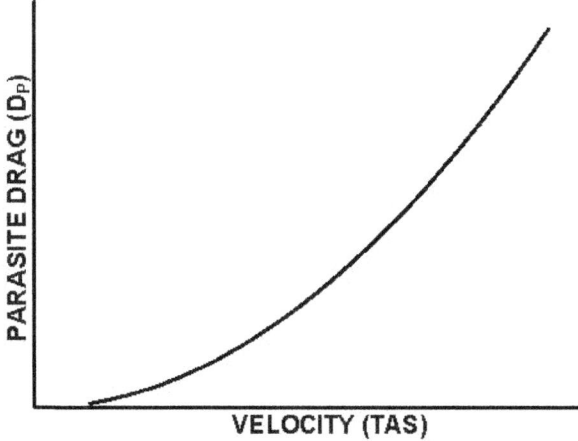

Figure 1-5-6 D$_P$ vs. Velocity

Total parasite drag (D$_P$) can be found by multiplying dynamic pressure by an area. **Equivalent parasite area (f)** is the area of a flat plate perpendicular to the relative wind that would produce the same amount of drag as form drag, friction drag and interference drag combined. It is not the cross-sectional area of the airplane. The equation for D$_P$ is:

$$D_P = \tfrac{1}{2}\rho V^2 f = qf$$

Parasite drag varies directly with velocity squared (V^2), so a doubling of speed will result in four times as much parasite drag (Figure 1-5-6).

INDUCED DRAG

INFINITE WING

Consider a wing placed in a wind tunnel with the tips flush against the walls. For all practical purposes it has no wingtips and is called an infinite wing. The relative wind on an infinite wing can only flow chordwise, and therefore produces lift. As the relative wind flows around the infinite wing, the high pressure air under the leading edge attempts to equalize

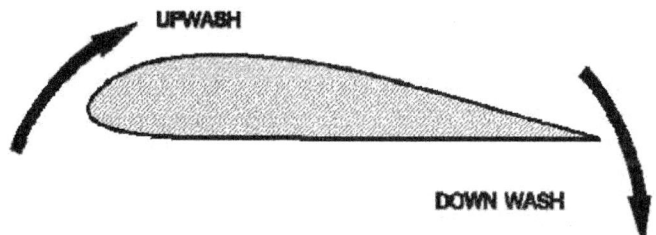

Figure 1-5-7 Upwash and Downwash

with the low pressure air above the wing. The shortest route is around the leading edge. This results in some of the air that otherwise would have passed under the wing flowing up and over the leading edge. This flow is called **upwash**. Upwash increases lift because it increases the average angle of attack on the wing. Some of the air on top of the wing also flows down and under the trailing edge. This flow is called **downwash**. Downwash decreases lift by reducing the average angle of attack on the wing. For an infinite wing, the upwash exactly balances the downwash resulting in no net change in lift. Upwash and downwash exist any time an airfoil produces lift.

FINITE WING

Upwash and downwash are not equal on a finite
wing. Not only does air flow up around the
leading edge of a finite wing producing upwash, it
also flows around the wingtips. Some of the high
pressure air in the leading edge stagnation point
flows spanwise to the wingtips instead of
chordwise over the upper surface of the wing.
Once it reaches the wingtips it flows around the
wingtips and up to the upper surface of the wing.
There, it combines with the chordwise flow that
has already produced lift and adds to the
downwash. Downwash approximately doubles

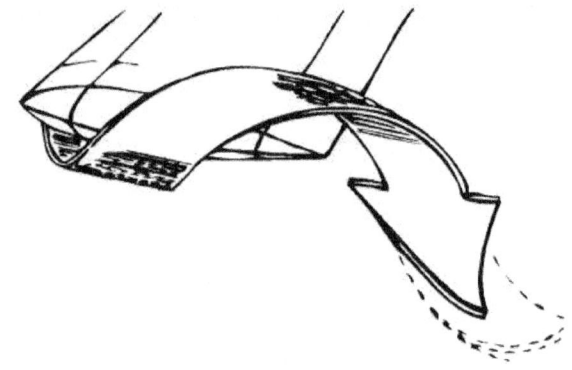

Figure 1-5-8 Finite Wing

by this process due to the spanwise airflow moving around the wingtip. The circular motion
imparted to the increased downwash also results in the formation of wingtip vortices.

Induced drag (D_i) is that portion of total drag associated with the production of lift. We can
add the airflow at the leading edge and the airflow at the trailing edge of the wing in order to
determine the average relative wind in the immediate vicinity of the wing. Since there is twice
as much downwash as upwash near the wingtips of a finite wing, the average relative wind has
a downward slant compared to the free airstream relative wind. The total lift vector will now be
inclined aft, as it in order to remain perpendicular to the average relative wind. The total lift
vector has components that are perpendicular and parallel to the free airstream relative wind.
The perpendicular component of total lift is called effective lift. Because total lift is inclined aft,
effective lift will be less than total lift. The parallel component of total lift is called induced drag
since it acts in the same direction as drag and tends to retard the forward motion of the
airplane.

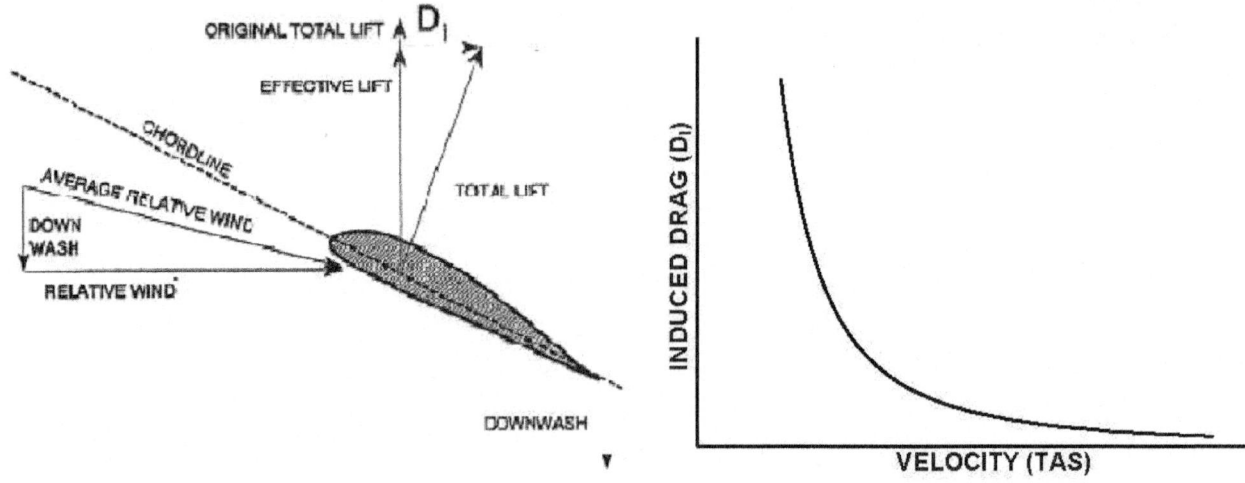

Figure 1-5-9 Induced Drag   Figure 1-5-10 D_i vs. Velocity

The D_i equation is derived from the aerodynamic force equation and the assumption that
weight equals lift in equilibrium level flight:

$$D_I = \frac{kL^2}{\rho V^2 b^2} = \frac{kW^2}{\rho V^2 b^2}$$

Analyzing the equation shows that increasing the weight of an airplane will increase induced drag, since a heavier airplane requires more lift to maintain level flight. Induced drag is reduced by increasing density (ρ), velocity (V), or wingspan (b). In level flight where lift is constant, induced drag varies inversely with velocity, and directly with angle of attack. Another method to reduce induced drag is to install devices that impede the spanwise airflow around the wingtip. These devices include winglets, wingtip tanks, and missile rails.

TOTAL DRAG

Parasite and Induced drag can be added together to create a total drag curve. By superimposing both drag curves on the same graph, and adding the values of induced and parasite drag at each velocity, the total drag curve of Figure 1-5-11 is derived. The numbers 1, 9, and 28 depicted near the curve are the angle of attack scale. Note that they decrease as TAS increases. The drag curve depicted is particular to one weight, one altitude and one configuration. As weight, altitude and configuration change, the total drag curve will shift.

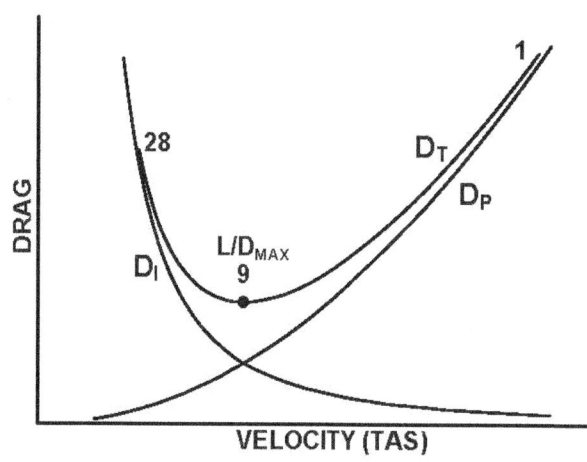

Figure 1-5-11 D_T vs. Velocity

LIFT TO DRAG RATIO

An airfoil is designed to produce lift, but drag is unavoidable. An airfoil that produced the desired lift but caused excessive drag would not be very useful. We use the lift to drag ratio (L/D) to determine the efficiency of an airfoil. A high L/D ratio indicates a more efficient airfoil. L/D is calculated by dividing lift by drag. All terms except C_L and C_D cancel out:

$$\frac{L}{D} = \frac{\frac{1}{2}\rho V^2 S C_L}{\frac{1}{2}\rho V^2 S C_D} = \frac{C_L}{C_D}$$

A ratio of the coefficients at a certain angle of attack determines the L/D ratio at that angle of attack. The L/D ratio can be plotted against angle of attack along with C_L and C_D (Figure 1-5-12). The maximum L/D ratio is called L/D_{MAX}. For the airplane in Figure 1-5-11 and Figure 1-5-12, L/D_{MAX} AOA is 9 units. Since angle of attack indicators are far less precise than airspeed indicators, pilots will typically fly an airspeed that corresponds to L/D_{MAX} AOA.

L/D_{MAX} AOA produces the minimum total drag. L/D_{MAX} is located at the bottom of the total drag curve. Any movement away from L/D_{MAX} will increase drag.

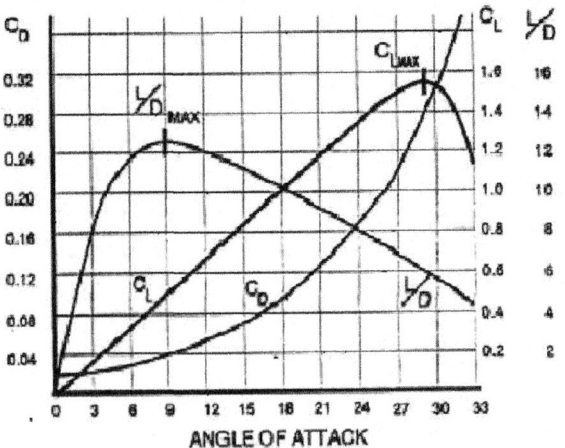

Figure 1-5-12 Lift to Drag Ratio

- At L/D_{MAX} AOA, parasite drag and induced drag are equal. At velocities below L/DMAX, the airplane is affected primarily by induced drag, while at velocities above L/D_{MAX}, the airplane is affected primarily by parasite drag.

- L/D_{MAX} AOA produces the greatest ratio of lift to drag. Note that this is not the maximum amount of lift that can be produced, nor does it correspond to the airplanes maximum speed.

- L/D_{MAX} AOA is the most efficient angle of attack. Note that L/D is the efficiency of the wing, not the engine.

- An increase in weight or altitude will increase L/D_{MAX} airspeed, but not affect L/D_{MAX} or L/D_{MAX} AOA. A change in configuration may have a large effect on L/D_{MAX} and L/D_{MAX} airspeed. The effect of configuration on L/D_{MAX} AOA will depend on what causes the change (lowering landing gear or flaps, dropping external stores, speed brakes, etc.), and how much change is produced. These changes in L/D_{MAX} are the topic of the next lesson.

1.   Define drag and list the factors that affect it.

2.   What happens to the coefficient of drag with an increase in angle of attack? When will the coefficient of drag equal zero?

3.   Define parasite drag and name the three primary types.

4.   State the equation for parasite drag, and identify its variables.

5.   State methods of reducing each type of parasite drag.

6.   Graph parasite drag vs velocity and describe their relationship.

7.   Describe the results of upwash and downwash around an infinite airfoil.

8.   Describe the airflow about a finite wing. What is the result of upwash and downwash on a finite wing?

9.   Define induced drag and describe how it is formed.

10.   State the equation for induced drag, and identify its variables.

11.   What can be done to reduce induced drag?

12.   Graph induced drag vs velocity and describe their relationship.

13.   State the equations for total drag.

14.   Increasing airspeed while maintaining level flight will cause (you may select more than one.):
        A.   An increase in induced drag
        B.   An increase in parasite drag
        C.   A decrease in induced drag
        D.   A decrease in parasite drag

15.   What does the L/D ratio represent? How is it calculated? How does a pilot determine what L/D ratio he is flying at?

16.   Define L/D_{MAX}. Locate L/D_{MAX} on the total drag curve. Why is it important?

Thrust and Power

INTRODUCTION

The purpose of this lesson is to aid the student in understanding thrust and thrust horsepower curves as they relate to aerodynamics.

TERMINAL OBJECTIVE

Upon completion of this unit of instruction, the student aviator will demonstrate knowledge of basic aerodynamic factors that affect airplane performance.

ENABLING OBJECTIVES

1.76	Describe the relationship between thrust and power.
1.77	Define thrust required and power required.
1.78	Describe how thrust required and power required vary with velocity.
1.79	State the location of L/D_{MAX} on the thrust required and power required curves.
1.80	Define thrust available and power available.
1.81	Describe the effects of throttle setting, velocity, and density on thrust available and power available for a turbojet engine.
1.82	Describe the effects of PCL setting, velocity, and density on thrust available and power available for a turboprop engine.
1.83	Define thrust horsepower.
1.84	Define shaft horsepower.
1.85	Define propeller efficiency.
1.86	State the relationship between thrust horsepower, shaft horsepower, and propeller efficiency.
1.87	State the flat rated shaft horsepower and the Navy limited shaft horsepower of the T-34C PT6A-25 engine.
1.88	State the instrument indications for the flat rated shaft horsepower and the Navy limited shaft horsepower of the T-34C PT6A-25 engine.
1.89	Define thrust excess and power excess.
1.90	State the effects of a thrust excess or a power excess.
1.91	State the conditions necessary to achieve the maximum thrust excess and maximum power excess for a turbojet and a turboprop airplane.
1.92	Describe the effects of changes in weight on thrust required, power required, thrust available, and power available.

1.93   Describe the effects of changes in weight on maximum thrust excess and maximum power excess, and on the airspeeds necessary to achieve maximum thrust excess and maximum power excess.

1.94   Describe the effects of changes in altitude on thrust required, power required, thrust available, and power available.

1.95   Describe the effects of changes in altitude on maximum thrust excess and maximum power excess, and on the airspeeds necessary to achieve maximum thrust excess and maximum power excess.

1.96   Describe the effects of changes in configuration on thrust required, power required, thrust available, and power available.

1.97   Describe the effects of changes in configuration on maximum thrust excess and maximum power excess, and on the airspeeds necessary to achieve maximum thrust excess and maximum power excess.

Thrust and Power

INTRODUCTION

The various aspects of airplane performance result from a combination of airframe and powerplant characteristics. These characteristics at various conditions of flight are depicted on thrust and power curves. These can be used to find maximum endurance, range, angle of climb, rate of climb, glide endurance and glide range.

REFERENCES

1.   Aerodynamics for Naval Aviators

2.   T-34C NATOPS Flight Manual

INFORMATION

THRUST AND POWER CURVES

In order to gain an understanding of various performance profiles, we must first establish some assumptions about our airplane and the charts we are about to use:

1.   Equilibrium flight on a standard day.

2.   No afterburner for a turbojet.

3.   Fixed pitch propeller for a turboprop.

THRUST REQUIRED

Recall that the total drag curve is the sum of parasite and induced drag. In equilibrium flight total thrust must equal total drag. Therefore, the amount of thrust that is required to overcome drag can be found on the total drag curve. This amount of thrust is called **thrust required (T_R)**, and is expressed in pounds. As with the drag curve, the thrust required curve is for one specific weight, altitude and configuration. L/D_{MAX} AOA is the point of minimum thrust required, and is obtained at some specific velocity. Flight at greater velocities requires a reduction in AOA (to maintain a constant lift/weight ratio) and an increase in thrust (to match the increase in parasite drag). Flight at lower velocities requires an increase in angle of attack and an increase in thrust (to match the increase in induced drag).

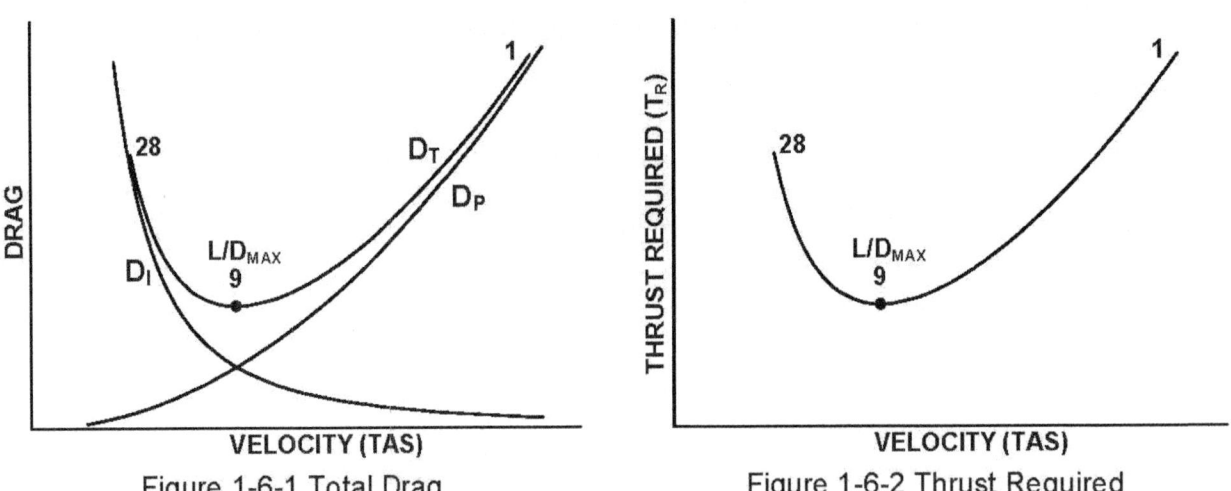

Figure 1-6-1 Total Drag   Figure 1-6-2 Thrust Required

POWER REQUIRED

Power is the rate of doing work, and work is a force times a distance. Power required (P_R) is the amount of power that is required to produce thrust required. P_R is the product of T_R and velocity (V). If V is expressed in knots, then the product of T_R and V must be divided by 325 to give power in units of horsepower. Thus, thrust horsepower only depends on thrust and velocity. For simplicity, we will use the term power (P) rather than thrust horsepower (THP) or shaft horsepower (SHP) unless there is a significant difference.

$$P_R = \frac{T_R \cdot V}{325}$$

To find L/D_{MAX} on the thrust required curve, draw a horizontal line tangent to the bottom of the curve. By applying the power equation to this line, the result is a straight line from the origin that is tangent to the power curve at L/D_{MAX}. Unlike on the T_R curve, L/D_{MAX} is not at the bottom of the P_R curve, but is to the right of the bottom of the curve. L/D_{MAX} still represents minimum total drag, but minimum P_R is to the left of L/D_{MAX}. It should be noted that the velocity and AOA for L/D_{MAX} are the same on the P_R curve as on the T_R curve.

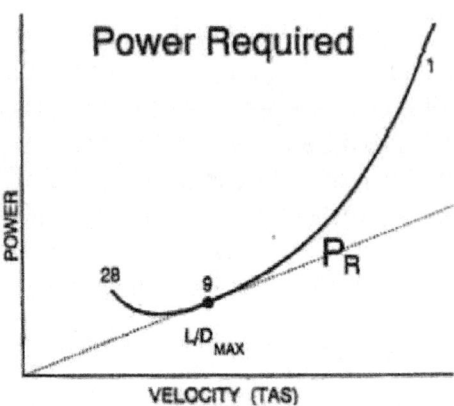

Figure 1-6-3 Power Required

THRUST AVAILABLE

Thrust available (T_A) is the amount of thrust that the airplane's engines actually produce at a given throttle setting, velocity, and density. The most important factor is the throttle, called the power control lever (PCL) in turboprops. For simplicity, we will use the term throttle. Maximum engine output occurs at full throttle. As the throttle is retarded, thrust available decreases. Since the propeller can only accelerate the air to a maximum velocity, as the velocity of the incoming air increases, the air is accelerated less through the propeller, and thrust available decreases (Figure 1-6-5).

Turbojets do not suffer a decrease in thrust available with velocity because ram-effect overcomes the decreased acceleration (Figure 1-6-4). Therefore, T_A is approximated by a straight line. As the density of the air decreases, thrust available decreases.

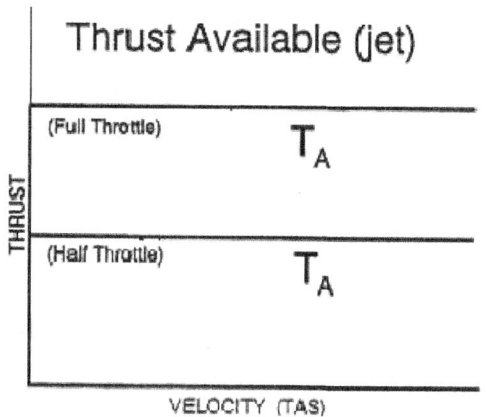

Figure 1-6-4 Thrust Available (Turbojet)

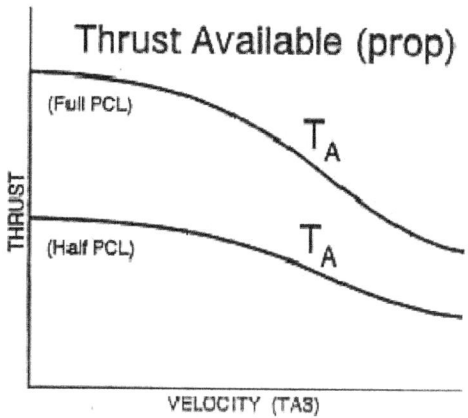

Figure 1-6-5 Thrust Available (Turboprop)

POWER AVAILABLE

Power available (P_A) is the amount of power that the airplane's engines actually produce at a given throttle setting, velocity, and density. The most important factor is throttle setting. Maximum power available occurs at full throttle. As the throttle is retarded, power available decreases. As velocity increases, power available for a jet will increase linearly, while power available for a prop will initially increase, but will then decrease due to a decrease in thrust available (Figure 1-6-6 and Figure 1-6-7). As thrust available decreases with a decrease in density, power available will also decrease. At sea level, the PT6A-25 engine in the T-34C is flat rated at 550 SHP (1315 ft-lbs of torque), but is Navy limited to 425 SHP (1015 ft-lbs of torque) in order to extend the service life of the engine.

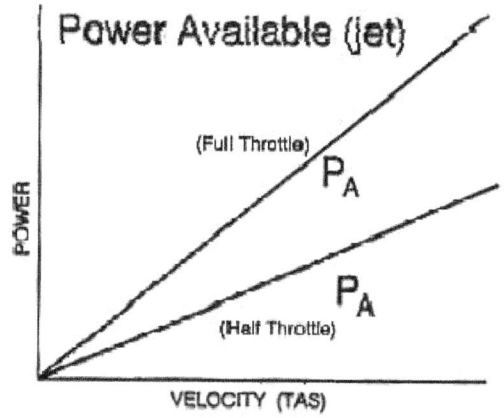
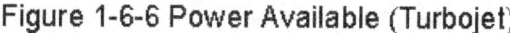

Figure 1-6-6 Power Available (Turbojet)

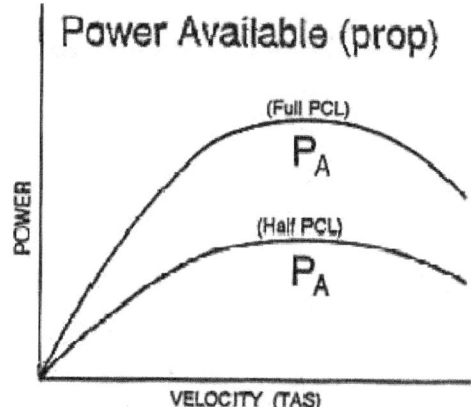

Figure 1-6-7 Power Available (Turboprop)

$$P_A = \frac{T_A \cdot V}{325}$$

In a turboprop, power available is determined by the performance of the engine/propeller combination. Engine output is called **shaft horsepower (SHP)**. **Thrust horsepower (THP)** is propeller output, or the power that is converted to usable thrust by the propeller. The ability of the propeller to turn engine output into thrust is given by its **propeller efficiency (p.e.)**. Under ideal conditions, SHP would equal THP, but due to friction in the gearbox and propeller drag, THP is always less than SHP. Propeller efficiency is always less than 100%.

$$THP = SHP \cdot p.e.$$

The T-34C has a constant-speed, variable-pitch propeller. Propeller efficiency will decrease as altitude increases. As the air density decreases, the blade angle will increase as the propeller takes a bigger "bite" of air to maintain a constant speed of 2200 rpm. At some point the blade angle will reach an optimum angle, beyond which efficiency will decrease with further in creases in blade angle. Note that a variable pitch, constant-speed propeller is more efficient than the fixed pitch propeller which is the basis of our thrust and power curves.

THRUST EXCESS AND POWER EXCESS

A comparison of the T_R and T_A curves on one graph allows us to predict airplane performance. To maintain equilibrium level flight, thrust available must equal thrust required for a specific angle of attack and velocity. This is depicted on a graph where the T_R and T_A curves cross. The right-hand point of equilibrium will produce the maximum velocity in level flight. This is the greatest airspeed that the aircraft can maintain without descending. It is approximately 190 KIAS at sea level for the T-34C.

A **thrust excess (T_E)** occurs if thrust available is greater than thrust required at a particular velocity. A positive TE causes an acceleration, a climb, or both, depending on angle of attack. A negative T_E is called a thrust deficit and has the opposite effect. Maximum thrust excess occurs at a full throttle setting, and is depicted on a graph where the distance between the T_R and T_A curves is greatest. For a turbojet, max thrust excess occurs at L/D_{MAX}. For a turboprop, max thrust excess occurs at a velocity less than L/D_{MAX}.

$$T_E = T_A - T_R$$

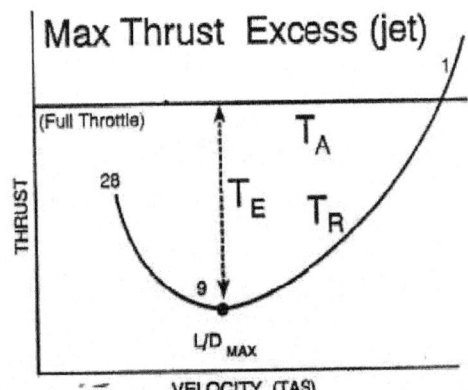
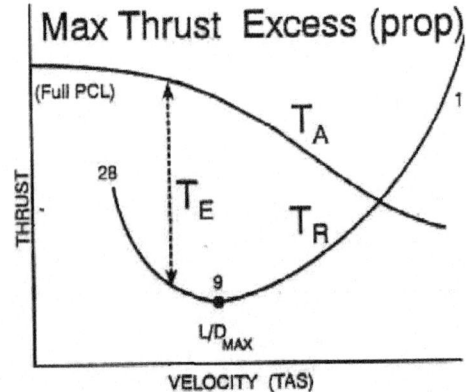

Figure 1-6-8 Thrust Excess (Turbojet) Figure 1-6-9 Thrust Excess (Turboprop)

Power excess (P_E) is calculated in a similar manner as T_E and will also produce an accelera-tion, a climb, or both. Likewise, a power deficit will cause a decent, a deceleration, or both. For a turbojet, maximum power excess occurs at a velocity greater than L/D_{MAX}. For a turbo-prop, max power excess occurs at L/D_{MAX}. It is important to note that maximum power excess is achieved at a greater velocity and a lower angle of attack than maximum thrust excess. It should also be noted that a power excess cannot exist if thrust excess is zero.

$$P_E = P_A - P_R$$

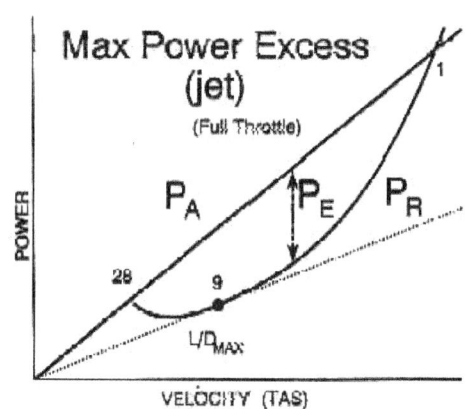

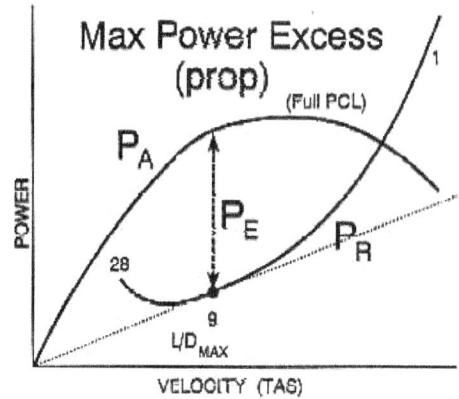

Figure 1-6-10 Power Excess (Turbojet)   Figure 1-6-11 Power Excess (Turboprop)

FACTORS AFFECTING T_E AND P_E

WEIGHT

If an airplane is in equilibrium level flight at a constant angle of attack, an increase in weight requires an increase in lift. In order to increase lift at a constant AOA, velocity must increase. This shifts the T_R curve to the right (Figure 1-6-12).

$$W = L = \tfrac{1}{2}\overline{\rho}\overset{\uparrow}{V^2}\overline{SC_L}$$

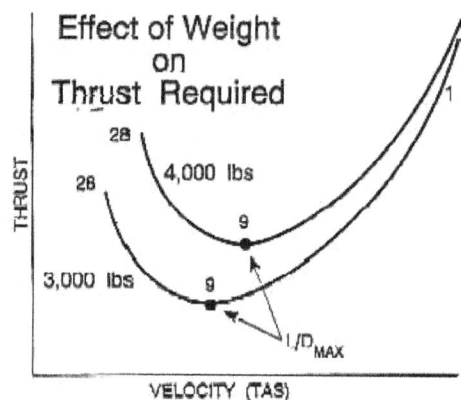

A higher velocity and more lift increase both parasite and induced drag, so total drag increases and the T_R shifts curve up (note that C_D remains constant if AOA is held constant).

Figure 1-6-12 Effect of Weight on T_R

$$T_R = D = \tfrac{1}{2}\overline{\rho}\overset{\uparrow}{V^2}\overline{SC_D}$$

Power required (P_R) is similarly affected by weight. An increase in weight requires an increase in velocity and a corresponding increase in thrust required (T_R) at a specific angle of attack. Since P_R is a function of thrust required and velocity, an increase in weight will result in an increase in power required. The net result of an increase in weight is that the T_R and P_R curves will shift up and right (Figure 1-6-13).

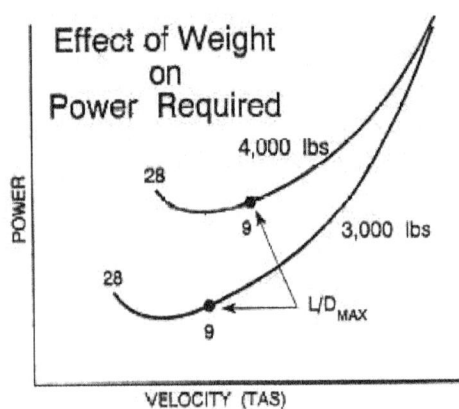

Figure 1-6-13 Effect of Weight on P_R

$$\overset{\uparrow}{P_R} = \frac{\overset{\uparrow}{T_R} \cdot \overset{\uparrow}{V}}{325}$$

Weight changes have no effect on thrust available or power available, as they do not affect the engine. As weight increases, thrust required and power required increase while thrust available and power available remain constant. Thus thrust excess and power excess decrease at every AOA and velocity.

$$\overset{\downarrow}{T_E} = \overline{\overset{}{T_A}} - \overset{\uparrow}{T_R}$$

$$\overset{\downarrow}{P_E} = \overline{\overset{}{P_A}} - \overset{\uparrow}{P_R}$$

ALTITUDE

If an airplane weighs 5,000 lbs at sea level, it requires 5,000 lbs of lift. It will weigh 5,000 lbs and require the same lift at any higher altitude as well. Since density has decreased, velocity must increase to maintain 5,000 lbs of lift. Thus as altitude increases, the T_R curve shifts to the right.

$$\overline{W} = \overline{L} = \tfrac{1}{2} \overset{\downarrow}{\rho} \overset{\uparrow}{V^2} \overline{S C_L}$$

Note that the decrease in density is exactly offset by an increase in velocity to maintain constant lift for any give AOA. In other words, the dynamic pressure felt by the airfoil remains constant.

With no change in dynamic pressure as lift is maintained at a higher altitude for any fixed AOA, drag and T_R remain constant. Thus as altitude increases, the thrust required curve shifts to the right, but not up (Figure 1-6-14).

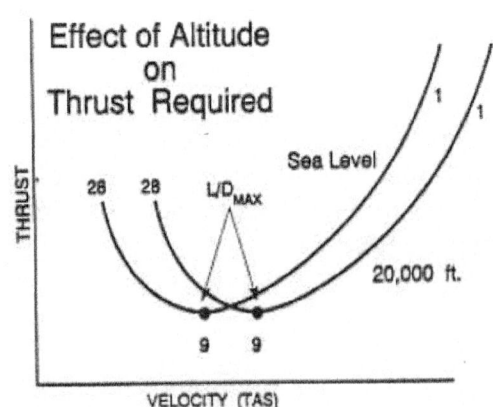

Figure 1-6-14 Effect of Altitude on T_R

$$\overline{T_R} = \overline{D} = \tfrac{1}{2} \overset{\downarrow}{\rho} \overset{\uparrow}{V^2} \overline{SC_D}$$

Since P_R is the product of T_R and velocity, the P_R curve will shift to the right as altitude increases and the T_R curve shifts to the right. However, since the same thrust is multiplied by a higher velocity, the P_R curve will move up as well (Figure 1-6-15). Both the T_R and P_R curves flatten slightly because of the decreasing effects of compressibility.

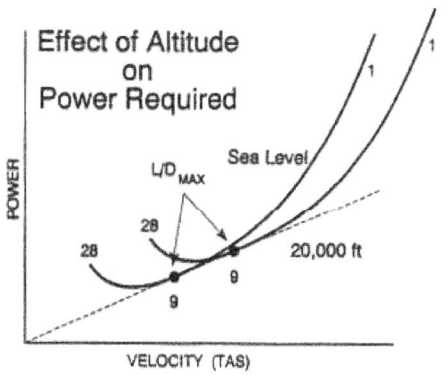

Figure 1-6-15 Effect of Altitude on P_R

$$\overset{\uparrow}{P_R} = \frac{\overline{T_R} \cdot \overset{\uparrow}{V}}{325}$$

Maximum engine output decreases with a reduction in air density. Thus, both T_A and P_A decrease at higher altitudes. Thrust excess will decrease with an increase in altitude due to the decrease in thrust available. Power excess will decrease with an increase in altitude because power available decreases and power required increases.

$$\overset{\downarrow}{T_E} = \overset{\downarrow}{T_A} - \overline{T_R}$$

$$\overset{\downarrow}{P_E} = \overset{\downarrow}{P_A} - \overset{\uparrow}{P_R}$$

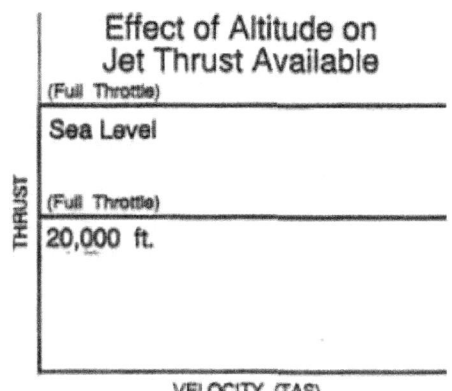

Figure 1-6-16 Effect of Altitude on T_A (Turbojet)

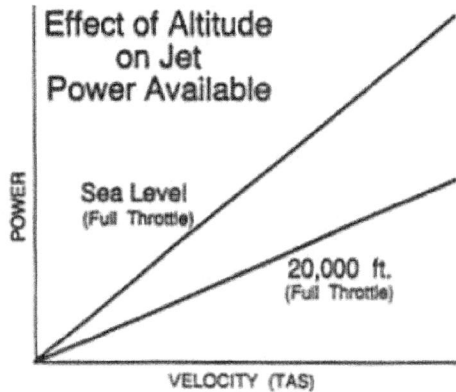

Figure 1-6-17 Effect of Altitude on P_A (Turbojet)

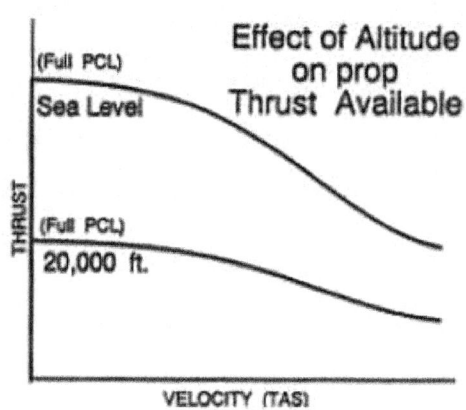

Figure 1-6-18 Effect of Altitude on T_A
(Turboprop)

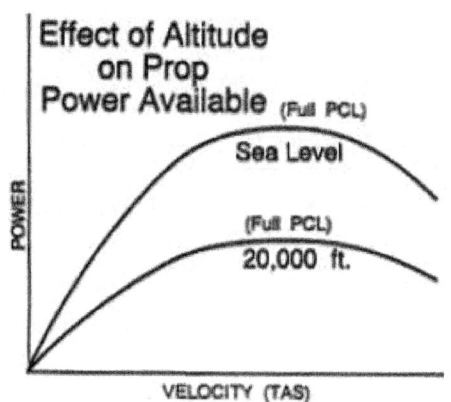

Figure 1-6-19 Effect of Altitude on P_A
(Turboprop)

CONFIGURATION

Lowering the landing gear has no effect on the lift produced by the wing, so at any AOA no change in velocity is required to maintain lift. Lowering the landing gear does, however, dramatically increases parasite drag, which causes T_R and P_R to increase. Thus more thrust and power are required to maintain altitude for any given AOA and velocity, so both the T_R and P_R curves shift up.

$$T_R = D = \tfrac{1}{2} \overset{\uparrow}{\overline{\rho V^2 S}} \overset{\uparrow}{C_D}$$

The landing gear has no effect on the engine, so T_A and P_A are not affected. Thrust and power excess will decrease with deployment of the landing gear because T_R and P_R increase.

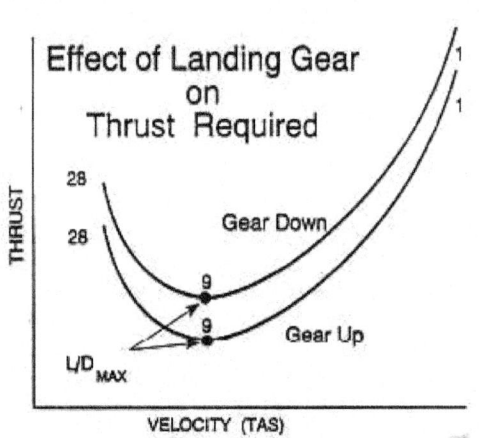

Figure 1-6-20 Effect of Landing Gear on T_R

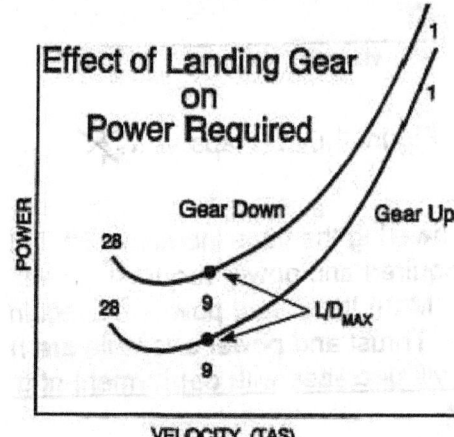

Figure 1-6-21 Effect of Landing Gear on P_R

Lowering the flaps increases the coefficient of lift, allowing the aircraft to fly at a lower velocity to produce enough lift to offset weight, so the T_R curve shifts left.

$$\overline{W} = \overline{L} = \tfrac{1}{2}\,\overline{\rho}\,\overset{\downarrow}{V^2}\,\overline{S}\,\overset{\uparrow}{C_L}$$

The flaps greatly increase parasite drag. Induced drag also increases.

Thus total drag and thrust required increase. Viewed another way using the drag equation, the decrease in velocity is more than offset by the increase in the coefficient of drag, causing thrust required to increase.

$$\overset{\uparrow}{T_R} = \overset{\uparrow}{D} = \tfrac{1}{2}\,\overline{\rho}\,\overset{\downarrow}{V^2}\,\overline{S}\,\overset{\uparrow\uparrow}{C_D}$$

The net effect of lowering flaps is to shift both the T_R and P_R curves up and to the left. More thrust and power are required to maintain altitude for any given velocity.

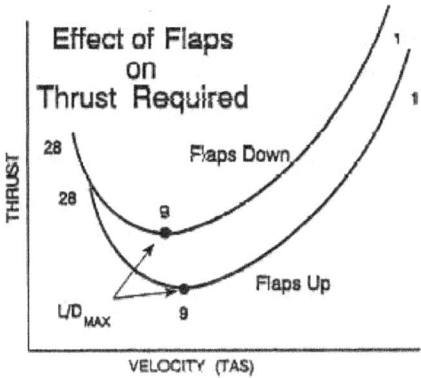
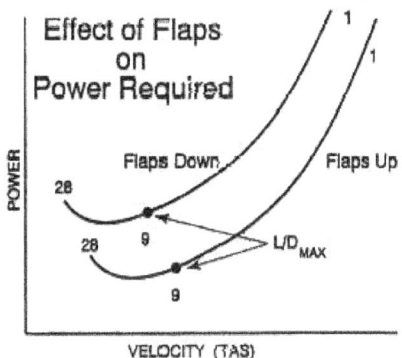

Figure 1-6-22 Effect of Flaps on T_R Figure 1-6-23 Effect of Altitude on P_R

As with the landing gear, flaps have no effect on the engine, so T_A and P_A are not affected. Thrust and power excess will decrease with deployment of the flaps because thrust and power required increase.

1.   Define thrust required. Where does the T_R curve come from?

2.   How is power computed from thrust?

3.   Minimum T_R occurs _____ L/D_{MAX} and minimum P_R occurs _____ L/D_{MAX}.
     A.   at, faster than
     B.   slower than, at
     C.   at, slower than
     D.   faster than, at

4.   Define thrust available and power available.

5.   The T-34C is powered by the Pratt & Whitney _____ turboprop engine. It can produce _____ SHP, but is Navy limited to _____ SHP at 2200 RPM.

6.   The angle of attack for maximum T_E will be _____ for a turbojet and _____ for a turboprop.
     A.   less than L/D_{MAX} AOA, equal to L/D_{MAX} AOA
     B.   equal to L/D_{MAX} AOA, less than L/D_{MAX} AOA
     C.   equal to L/D_{MAX} AOA, greater than L/D_{MAX} AOA
     D.   greater than L/D_{MAX} AOA, equal to L/D_{MAX} AOA

7.   The airspeed for max P_E will be _____ for a turbojet and _____ for a turboprop.
     A.   less than L/D_{MAX}, equal to L/D_{MAX}
     B.   equal to L/D_{MAX}, less than L/D_{MAX}
     C.   equal to L/D_{MAX}, greater than L/D_{MAX}
     D.   greater than L/D_{MAX}, equal to L/D_{MAX}

8.   How are the T_R and P_R curves affected by increased weight? Increased altitude?

9.   How are T_A and P_A affected by throttle/PCL setting? By altitude? By weight?

10.   How do altitude and weight affect P_E and T_E?

Airplane Performance

INTRODUCTION

The purpose of this lesson is to aid the student in understanding thrust and power as they relate to aircraft performance.

TERMINAL OBJECTIVE

Upon completion of this unit of instruction, the student aviator will demonstrate knowledge of basic aerodynamic factors that affect airplane performance.

ENABLING OBJECTIVES

1.98	State the relationship between fuel flow, thrust available, thrust required, and velocity for a turbojet airplane in straight and level flight.
1.99	State the relationship between fuel flow, power available, power required, and velocity for a turboprop airplane in straight and level flight.
1.100	Define maximum endurance and maximum range.
1.101	State the angle of attack and velocity, compared to L/D_{MAX}, at which turbojet and turboprop airplanes achieve maximum endurance.
1.102	State the angle of attack and velocity, compared to L/D_{MAX}, at which turbojet and turboprop airplanes achieve maximum range.
1.103	Describe the effect of changes in weight, altitude, configuration, and wind on maximum endurance and maximum range performance and airspeed.
1.104	Define maximum angle of climb and maximum rate of climb.
1.105	State the angle of attack and velocity, compared to L/D_{MAX}, at which turbojet and turboprop airplanes achieve maximum angle of climb.
1.106	State the angle of attack and velocity, compared to L/D_{MAX}, at which turbojet and turboprop airplanes achieve maximum rate of climb.
1.107	Describe the effect of changes in weight, altitude, configuration, and wind on maximum angle of climb and maximum rate of climb performance and airspeed.
1.108	Define absolute ceiling, service ceiling, cruise ceiling, combat ceiling, and maximum operating ceiling.
1.109	State the maximum operating ceiling of the T-34C.
1.110	Define maximum glide range and maximum glide endurance.
1.111	State the angle of attack and velocity, compared to L/D_{MAX}, at which an airplane achieves maximum glide range.
1.112	State the angle of attack and velocity, compared to L/D_{MAX}, at which an airplane achieves maximum glide endurance.

1.113   Describe the effects of changes in weight, altitude, configuration, wind, and propeller feathering on maximum glide range and maximum glide endurance performance and airspeed.

1.114   Define the regions of normal and reverse command as they relate to maximum endurance angle of attack and velocity.

1.115   Describe the relationship between velocity and throttle setting required to maintain level flight within the region of normal and reverse command.

Airplane Performance

INTRODUCTION

The thrust and power curves allow one to find maximum endurance, range, angle of climb, rate of climb, glide endurance and glide range. This lesson covers the basic concepts in this area.

REFERENCES

1.   Aerodynamics for Naval Aviators

2.   Aerodynamics for Pilots

3.   Introduction to the Aerodynamics of Flight

4.   T-34C NATOPS Flight Manual

INFORMATION

LEVEL FLIGHT PERFORMANCE

Fuel flow is the rate of fuel consumption by the engine, measured in pounds per hour (pph). Since the supply of fuel onboard is limited, the engine's fuel flow is a critical determinant of how long and how far the airplane can fly. A turbojet engine directly produces thrust through its exhaust. Therefore, the fuel consumed by a turbojet engine is proportional to its thrust available (T_A). In order to maintain equilibrium flight, thrust available must be set equal to thrust required (T_R), therefore we say that minimum fuel flow for a turbojet is found on the thrust required curve.

The thrust provided by a propeller is not produced directly by the engine, so there is no direct relationship between thrust and fuel flow. The engine turns a shaft that turns the propeller that produces the thrust. In turning the shaft, the engine produces power. Therefore, for a turbo-prop, fuel flow varies directly with the power output of the engine (P_A). However, minimum fuel flow for equilibrium flight will be found on the power required (P_R) curve.

Maximum endurance and maximum range are both achieved in equilibrium, level flight. Any thrust or power excess would cause the airplane to either climb or accelerate. We will look on the thrust required or power required curve to determine the velocity that our airplane must fly. Once the velocity is determined, the pilot must adjust the throttle to eliminate any thrust or power excess.

Maximum endurance is the maximum amount of time that an airplane can remain airborne on a given amount of fuel. The slower an engine burns fuel, the longer the airplane can remain airborne. Minimum fuel flow occurs at minimum T_R for a turbojet and minimum P_R for a turbo-prop. Therefore, maximum endurance is found at L/D_{MAX} AOA and velocity for a turbojet and at a velocity less than L/D_{MAX}, and an angle of attack greater than L/D_{MAX} AOA for a turboprop. For the T-34C, maximum endurance is achieved at approximately 420 ft-lbs of torque.

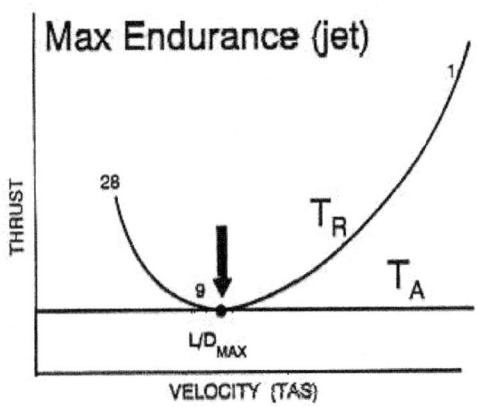

Figure 1-7-1 Turbojet Maximum Endurance

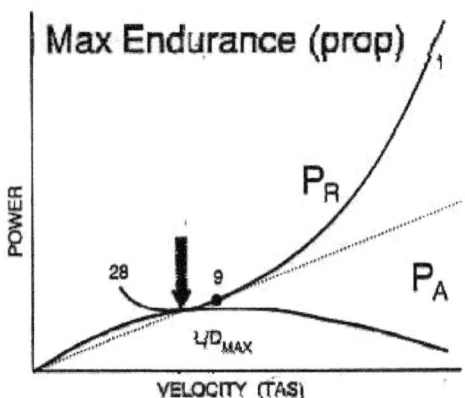

Figure 1-7-2 Turboprop Maximum Endurance

Maximum range is the maximum distance traveled over the ground for a given amount of fuel. To find maximum range we must minimize fuel flow per unit of velocity. Any straight line drawn from the origin represents a constant ratio of fuel flow to velocity. The minimum ratio that allows the airplane to remain airborne occurs where the line from the origin is tangent to the T_R curve for jets or the P_R curve for props. Maximum range for a turbojet is found at a velocity greater than L/D_{MAX} and an angle of attack less than L/D_{MAX} AOA. Maximum range for a turboprop is found at L/D_{MAX} AOA and velocity. Maximum range with no wind is achieved in the T-34C at approximately 580 ft-lbs of torque. Note that maximum range is faster than maximum endurance.

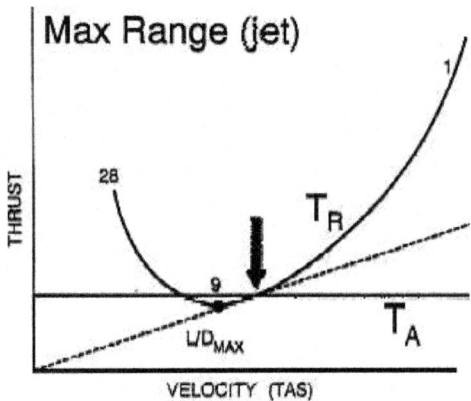

Figure 1-7-3 Turbojet Maximum Range

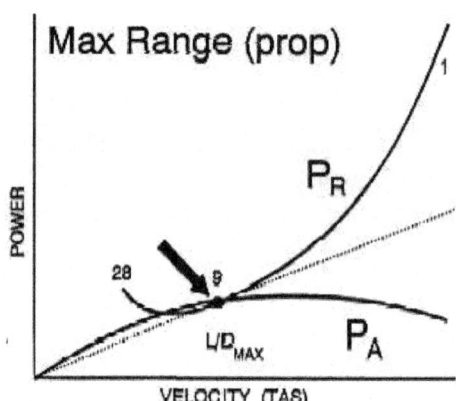

Figure 1-7-4 Turboprop Maximum Range

As explained in the previous lesson, if the weight of an airplane increases, the thrust required curve (Figure 1-6-12) and the power required curve (Figure 1-6-13) will both shift to the right and up. The shift to the right is due to the higher velocity required to produce more lift. Since thrust represents fuel flow for a turbojet, as T_R increases so will fuel flow for a turbojet. As P_R increases, fuel flow for a turboprop will increase. Higher fuel flow means maximum endurance performance will decrease with an increase in weight and max endurance airspeed will increases. The increased fuel flow will also decrease maximum range performance and increase max range airspeed.

An increase in altitude moves the thrust required curve to the right (Figure 1-6-14) and the power required curve to the right and up (Figure 1-6-15). However, as altitude increases (sea level to 36,000 ft MSL), the temperature rapidly decreases (to –56.5°C). Decreased temperatures make turbine engines more fuel efficient, requiring less fuel for a given amount of thrust or power. Although the pilot physically increases the throttle setting as altitude increases, fuel flow *decreases*. Since the airplane is burning less fuel to remain airborne, maximum endurance performance increases with an increase in altitude.

An airplane at a higher altitude will fly at a greater TAS while burning less fuel. Since the fuel consumed per mile flown has decreases, an increase in altitude increases maximum range performance. With the same increase in altitude, turbojet airplane will notice a greater gain in performance than turboprop airplane. This is due in part to the loss of propeller efficiency with altitude. Early turboprop airplane experienced a loss of performance with an increase in altitude.

Configuration changes will affect both max endurance and max range. Lowering the landing gear or flaps causes the thrust required and power required curves to shift up (Figure 1-6-20 through Figure 1-6-23). Max endurance and max range will decrease with landing gear and/or flaps extended.

Since range is distance over the ground, ground speed must be considered when determining the effect of wind on maximum range. When we fly into a headwind, our ground speed is less than our true airspeed. Therefore, the range of the airplane decreases since less ground will be covered in a given time. Headwinds will decrease maximum range performance while tailwinds will increase maximum range performance. Winds will have no effect on maximum endurance performance.

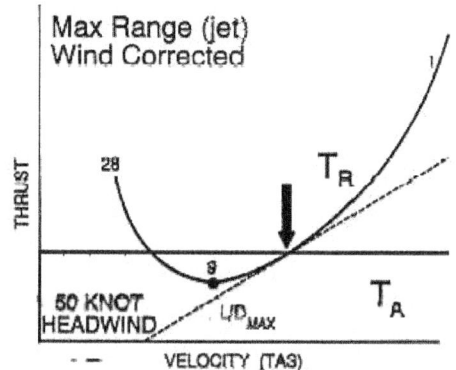
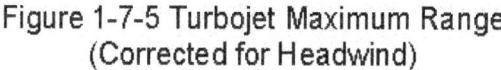
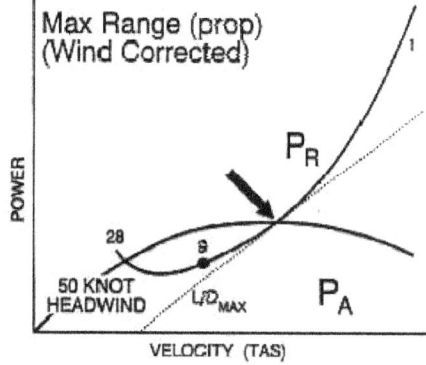

Figure 1-7-5 Turbojet Maximum Range   Figure 1-7-6 Turboprop Maximum Range
       (Corrected for Headwind)   (Corrected for Headwind)

To partially make up for the decreased performance with a headwind, we can increase the TAS of the airplane. We cannot totally make up for the distance lost, but some of the headwind effect can be overcome. Consider the extreme case of an airplane flying into a headwind that equals TAS. Ground speed and range are zero. *Any* increase in true airspeed would increase range. The straight line drawn from the origin tangent to the T_R or P_R curve represents a ratio of fuel flow to *true airspeed*. To make the tangent line represent a ratio of fuel flow to *ground speed*, we must subtract headwind or add a tailwind to true airspeed. With a headwind, we move the beginning of the tangent to the right of the origin by the amount of the headwind velocity (Figure 1-7-5 and Figure 1-7-6). The airspeed under the new tangent point is the velocity needed to fly maximum range with the headwind. With a tailwind, we move the beginning of the tangent to the left of the origin by the amount of the tailwind velocity.

CLIMB PERFORMANCE

A "steady climb" is defined as a climb in which the airplane is not accelerating; the airplane is in equilibrium. However, the altitude is no longer constant. In this discussion, the same thrust and power curves are used to analyze level flight to discuss and locate the different climb performance parameters of an airplane.

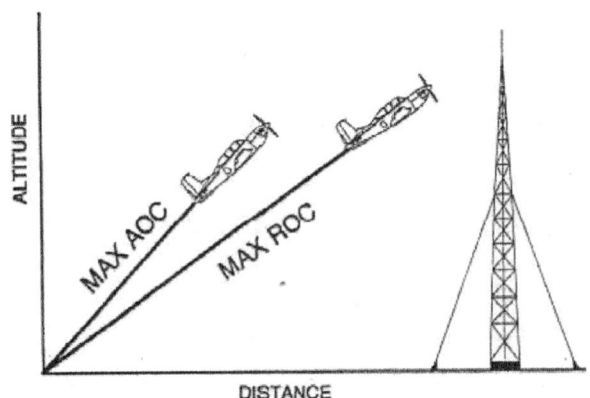

Figure 1-7-7 Max AOC vs. Max ROC

Angle of climb (γ, AOC) is a comparison of altitude gained to distance traveled. For maximum angle of climb, we want maximum vertical velocity (altitude increase) for a minimum horizontal velocity (ground speed). Maximum AOC is commonly used when taking off from a short airfield surrounded by high obstacles, such as trees, or power lines. The objective is to gain sufficient altitude to clear the obstacle with the least horizontal distance traveled. **Rate of climb (ROC)** is a comparison of altitude gained relative to the time needed to reach that altitude. Flying at maximum rate of climb yields a maximum vertical velocity. Maximum rate of climb is used to expedite a climb to an assigned altitude. The greatest vertical distance must be gained in the shortest time possible.

In a maximum angle of climb profile, a certain airplane takes 30 seconds to reach 1000 feet AGL, but covers only 3000 feet over the ground. Using its maximum rate of climb profile, the same airplane climbs to 1500 feet in 30 seconds, but covers 6000 feet across the ground. It should be noted that both climb profiles are executed at maximum throttle setting, and that differences between max rate and max angle of climb lie solely in differences of angle of attack and velocity.

ANGLE OF CLIMB

The equations that represent equilibrium in a climb are:

$$L = W \cos \gamma$$

$$T = D + W \sin \gamma$$

By rearranging the bottom equation, we see that:

$$\sin \gamma = \frac{T - D}{W} = \frac{T_A - T_R}{W} = \frac{T_E}{W}$$

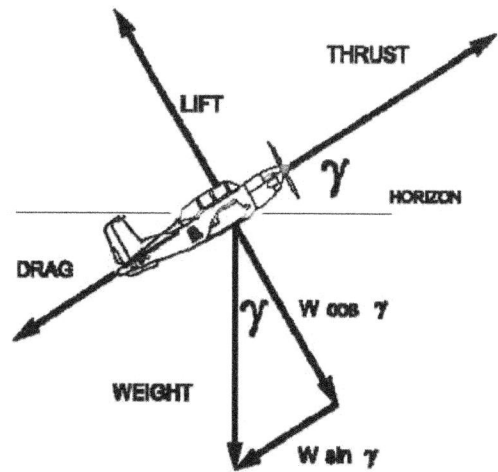

Figure 1-7-8 Climb Forces

Thus, angle of climb performance depends upon thrust excess. Essentially, the greater the force that pushes the airplane upwards, the steeper it can climb. Maximum angle of climb occurs at the velocity and angle of attack that produce the maximum thrust excess. Therefore, maximum angle of climb for a turbojet occurs at L/D_{MAX} AOA and velocity. Maximum angle of climb for a turboprop occurs at a velocity less than L/D_{MAX} and an angle of attack greater than L/D_{MAX} AOA. Maximum angle of climb airspeed (V_X) is approximately 75 KIAS for the T-34C.

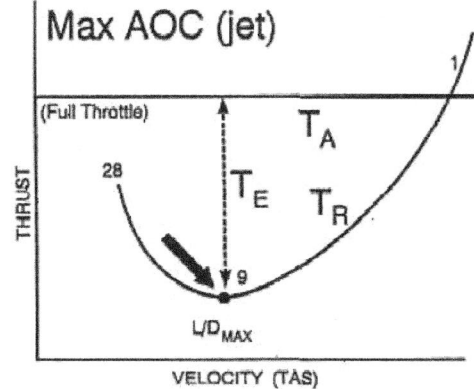

Figure 1-7-9 Turbojet Angle of Climb

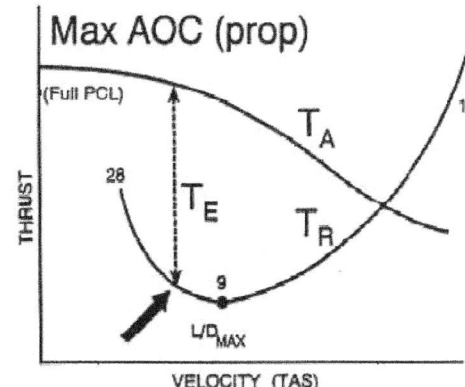

Figure 1-7-10 Turboprop Angle of Climb

RATE OF CLIMB

Rate of climb (ROC) is simply the vertical component of velocity (Figure 1-7-11):

$$ROC = V \sin \gamma$$

$$\sin \gamma = \frac{T_A - T_R}{W} = \frac{T_E}{W}$$

By substitution:

$$ROC = V \sin \gamma = \frac{VT_E}{W} = \frac{P_E}{W}$$

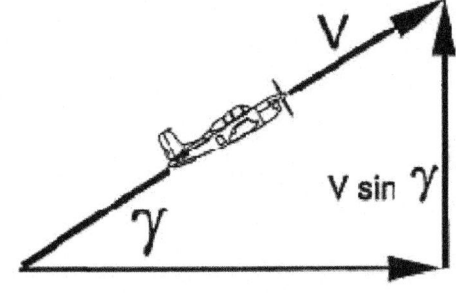

Figure 1-7-11 Climb Velocity Vectors

Thus, rate of climb performance depends upon power excess. Since climbing is work and power is the rate of doing work, any power that is not used to maintain level flight can increase the rate of climbing. Maximum rate of climb occurs at the velocity and angle of attack that produce the maximum power excess. Therefore, maximum rate of climb for a turbojet occurs at a velocity greater than L/D$_{MAX}$ and an angle of attack less than L/D$_{MAX}$ AOA. Maximum rate of climb for a turboprop occurs at L/D$_{MAX}$ AOA and velocity. Maximum rate of climb airspeed (V$_Y$) is approximately 100 KIAS for the T-34C.

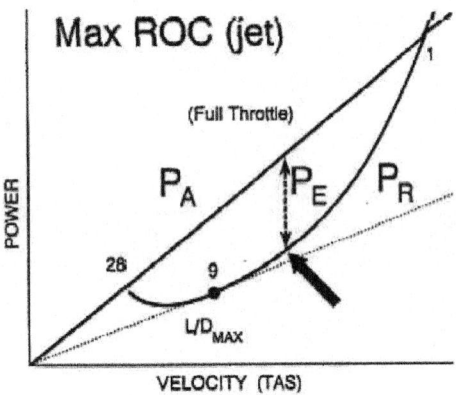

Figure 1-7-12 Turbojet Rate of Climb

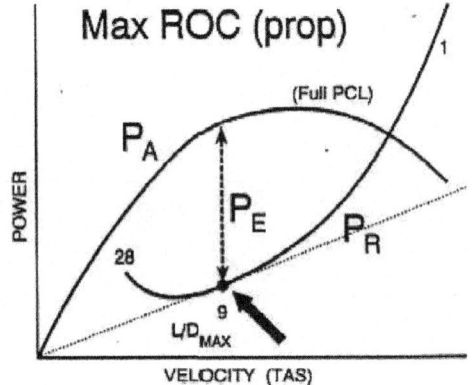

Figure 1-7-13 Turboprop Rate of Climb

CLIMB PERFORMANCE FACTORS

Since weight, altitude, and configuration changes affect thrust and power excess, they will also affect climb performance. Climb performance is directly dependent upon the ability to produce either a thrust excess or a power excess. In the previous lesson, it was determined that an increase in weight, an increase in altitude, lowering the landing gear, or lowering the flaps will all decrease both maximum thrust excess and maximum power excess for all airplane. Therefore, maximum angle of climb and maximum rate of climb performance will decrease under any of these conditions.

Consider an airplane that has a maximum
angle of climb TAS of 75 knots, a ground
speed of 60 knots, and no wind. If this
airplane flies into a headwind of 20 knots, its
ground speed is reduced to 40 knots. The
headwind has increased the airplane's
maximum angle of climb, because it reaches
the same altitude as before with a smaller
distance covered over the ground. A tailwind
has the opposite effect (Figure 1-7-14). Wind
does not affect rate of climb performance.

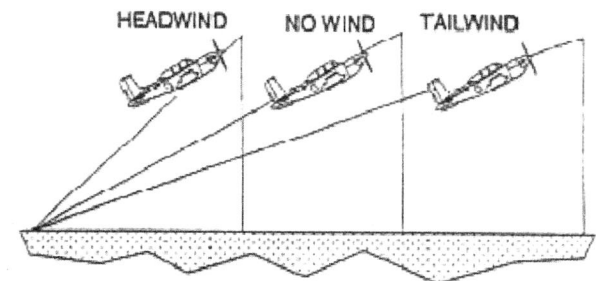

Figure 1-7-14 Effect of Wind on Max AOC

CEILINGS

As an airplane climbs and P_E decreases, the rate of climb will also decrease. The altitude
where maximum power excess allows only 500 feet per minute rate of climb is called the **com-
bat ceiling**. The **cruise ceiling** is the altitude at which an airplane can maintain a maximum
climb rate of only 300 feet per minute. The **service ceiling** is the altitude at which an airplane
can maintain a maximum rate of climb of only 100 feet per minute. Eventually, the airplane will
reach an altitude where maximum power excess is zero. At this altitude, the airplane can no
longer perform a steady climb, and its maximum rate of climb is zero. The altitude at which
this occurs is called the **absolute ceiling**. If the airplane flies at its maximum rate of climb
velocity, it will only be possible to maintain level equilibrium flight. At any velocity other than
this, P_R will exceed P_A, and the airplane will descend. The T-34C is limited to a maximum
operating ceiling of 25,000 feet. Federal Aviation Regulations prohibit unpressurized airplane
from exceeding 25,000 feet unless all occupants are wearing full pressure suits.

GLIDE PERFORMANCE

Gliding is a condition of flight without any operating
engine. It does not refer to a single engine failure in a
multi-engine airplane. When our engine fails, we may
need to glide as far as possible to reach a safe landing
area. This is a **maximum glide range** profile. If we lose
power within easy reach of a safe runway, we may decide
to fly a maximum glide endurance profile while the
runway is being cleared. The equations that represent
equilibrium in a glide are:

$$L = W \cos \gamma$$

$$D = W \sin \gamma$$

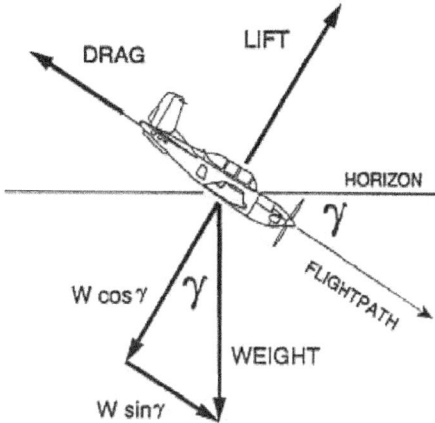

Figure 1-7-15 Glide Forces

GLIDE RANGE

To achieve maximum glide range, a pilot should maintain the minimum glide angle. Rearranging the above equation:

$$\sin \gamma = \frac{D}{W} = \frac{T_R - T_A}{W} = \frac{T_D}{W}$$

Thus, the angle of descent is directly related to the thrust deficit, T_D. To achieve the minimum angle of descent, we must minimize the thrust deficit, which occurs at L/D_{MAX} (Figure 1-7-16). Therefore, maximum glide range occurs at L/D_{MAX}. Maximum glide range velocity (V_{BEST}) is L/D_{MAX} for any airplane regardless of engine type. Since the L/D ratio is determined by angle of attack, any change away from L/D_{MAX} AOA would result in a decreased L/D ratio and a decrease in glide range. By holding a constant AOA, we can maintain a constant L/D ratio, regardless of weight or velocity. V_{BEST} is 100 KIAS for the T-34C. Glide range is often expressed as a ratio of horizontal distance to vertical distance. A glide ratio of 12:1 indicates that an airplane will move forward 12 feet for every foot of altitude lost.

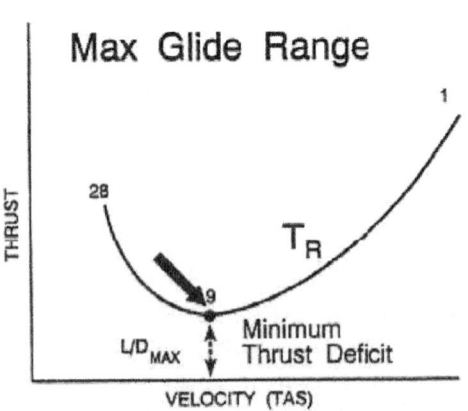

Figure 1-7-16 Max Glide Range

GLIDE ENDURANCE

Maximizing glide endurance is simply a matter of minimizing rate of descent (ROD) or negative vertical velocity.

$$ROD = V \sin \gamma$$

$$\sin \gamma = \frac{D}{W} = \frac{T_D}{W}$$

By substituting:

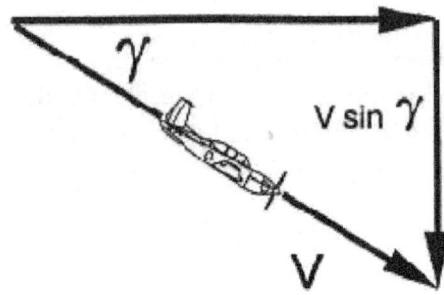

Figure 1-7-17 Glide Velocity Vectors

$$ROD = V \sin \gamma = \frac{V T_D}{W} = \frac{P_D}{W}$$

To minimize the rate of descent, the pilot must fly at the velocity where the minimum power deficit occurs. This is at the bottom of the P_R curve (Figure 1-7-18). Maximum glide endurance velocity is less than L/D_{MAX} velocity, and the angle of attack for max glide endurance is greater than L/D_{MAX} AOA. Maximum glide endurance is found at 87 KIAS in the T-34C. To provide for an adequate safety margin from stalls in the T-34 during power-off flight, 100 KIAS will always be used in the emergency landing pattern. This airspeed will also provide performance that is close to maximum glide endurance. Flying a consistent pattern is far more important than gaining an extra few seconds of glide endurance.

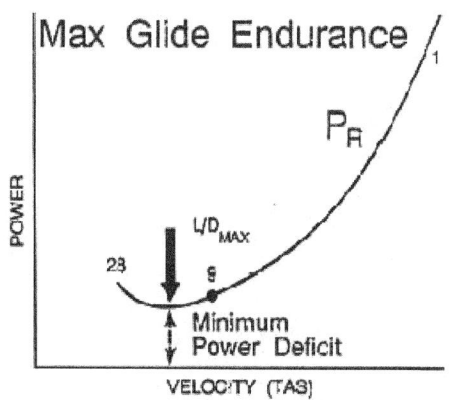

Figure 1-7-18 Max Glide Endurance

GLIDE PERFORMANCE FACTORS

One might feel a tendency to try to "stretch out" the glide by increasing the angle of attack. If the angle of attack is increased beyond L/D_{MAX} AOA, the horizontal distance the plane will travel will actually decrease. The minimum glide angle obtained at L/D_{MAX} will not produce the minimum sink rate, but will produce the greatest horizontal distance for a given altitude.

As the airplane's weight is increased, the T_R and P_R curves shift up and to the right. The lowest point on each curve will shift as well, increasing the velocity at which it occurs. As long as the pilot maintains L/D_{MAX} AOA, the L/D ratio and angle of descent remain constant. Therefore, an increase in weight will not affect maximum glide range. An increase in the velocity during a descent will cause the rate of descent to increase, and glide endurance to decrease. Increasing the weight will cause the airplane to fly faster and descend faster, but still glide the same distance.

An increase in altitude will increase the maximum glide range and maximum glide endurance of an airplane.

Wind has the same effect on maximum glide range that it has on maximum range. Since a headwind decreases groundspeed, it causes a decrease in the maximum glide range. Conversely, a tailwind will increase the maximum glide range. Wind has no effect on rate of descent or on glide endurance.

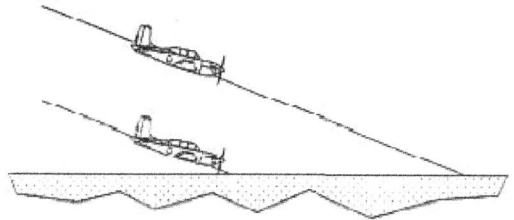

Figure 1-7-19 Effect of Altitude on Glide

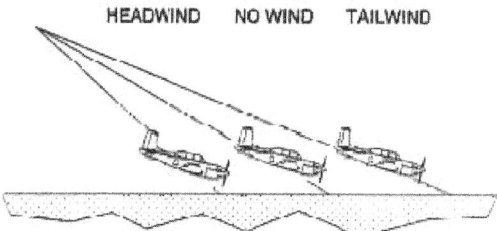

Figure 1-7-20 Effect of Altitude on Glide

During power off flight, airplane configuration plays a major role in determining glide performance. If the pilot alters the configuration by opening the canopy or extending the landing gear and/or flaps, the sink rate will increase and glide range will decrease. Some specific sink rates and glide ratios from theT-34C NATOPS Manual are listed below. Knowledge of these sink rates should assist the student in understanding the practice/actual emergency landing patterns and assist in achieving the proper position in them (Table 1-7-3).

The greatest effect of configuration on glide performance deals with the propeller. In normal flight, the propeller blades are almost flat to the relative wind, but create no drag since the engine is driving the prop around. When the engine fails, if the propeller blades stay flat to the relative wind, the wind will drive the propeller blades around, a situation called **windmilling**. Windmilling significantly increases the drag on the airplane and adversely affects glide performance. In order to stop the propeller from windmilling, the individual propeller blades can be turned so they are aligned with the wind. This procedure is called **feathering** the propeller. In Table 1-7-3, the importance of feathering the propeller in an actual engine failure is clear.

Configuration		Prop Feathered	Sink rate at 100 KIAS	Glide ratio at 100 KIAS
Gear	Flaps			
Up	Up	Yes	800 fpm	12 to 1
Down	Up	Yes	1,200 fpm	8 to 1
Up	Down	Yes	1,250 fpm	8 to 1
Down	Down	Yes	1,650 fpm	6 to 1
Up	Up	No	2,400 fpm	3 to 1

Table 1-7-3 T-34C Sink Rate Comparison

Note: Configurations shown include canopy closed, wings level and balanced flight. With both canopies open, the sink rate typically increases by 300 fpm.

THE REGIONS OF NORMAL AND REVERSE COMMAND

Velocities above maximum endurance are referred to as the **region of normal command**. The region of normal command is characterized by airspeed stability. Assume an airplane is in equilibrium at point B (Figure 1-7-21 or Figure 1-7-22). A decrease in airspeed (for whatever reason) results in a thrust or power excess that will eventually accelerate the airplane back to the original airspeed at point B. An increase in airspeed from point B (for whatever reason) results in a thrust or power deficit that slows the airplane back to the original airspeed.

In the region of normal command, velocity and throttle setting for level flight are directly related. To fly in equilibrium at a faster airspeed, more T_A/P_A is needed than at a slower airspeed. To fly slower, less T_A/P_A is needed.

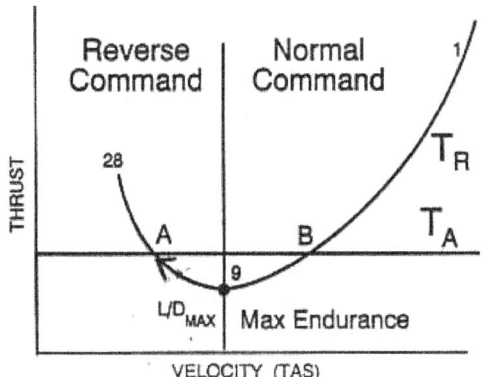

Figure 1-7-21 Turbojet Reverse Command

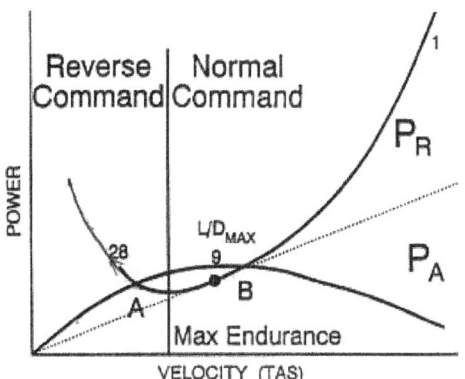

Figure 1-7-22 Turboprop Reverse Command

Velocities below maximum endurance are referred to as the **region of reverse command**.
The region of reverse command is characterized by airspeed instability. Assume an airplane is
in equilibrium at point A (Figure 1-7-21 or Figure 1-7-22). A decrease in airspeed (for whatever
reason) results in a thrust or power deficit that will eventually slow the airplane to the point of
stalling (assuming a level flight attitude is being maintained). An increase in airspeed (for
whatever reason) from Point A results in a thrust or power excess that accelerates the airplane
away from point A. The airplane eventually reaching equilibrium at point B.

In the region of reverse command, velocity and throttle setting for level flight are inversely
related. Once stabilized at a faster airspeed in equilibrium flight, T_A/P_A will be lower than when
stabilized at a slower airspeed. Simply stated, the slower an airplane flies in the region of
reverse command, the more thrust and power is needed.

A complete knowledge of this flight region is particularly important because most aviation
accidents occur while operating in the region of reverse command. Whenever an airplane is
taking off or landing, it is flying in or near this region. A very dangerous situation for an
inexperienced pilot is trying to slow down in the region of reverse command. If the pilot
increases back pressure to increase angle of attack and decrease velocity, this will causes
thrust and power required to increase, creating a deficit. Once the airspeed bleeds off, the
deficit causes the airplane to descend. The inexperienced pilot tends to pull back on the
control stick in order to keep from descending. This causes the airplane to move further into
the region of reverse command, creating a greater deficit. Eventually the deficit will be so
great that even full throttle is not able to overcome it. Since this usually occurs during landing,
there is not enough altitude to recover. This is the origin of the phrase "behind the power
curve." An experienced pilot knows that in order to maintain level flight as an airplane slows
down in the region of reverse command, the throttle must be increased. Increasing angle of
attack will only aggravate the situation.

1.   What curves determine fuel flow for a turboprop and a turbojet?

2.   Define maximum endurance and maximum range.

3.   What performance profiles occur at L/D_{MAX}?
     A.   Jet max range, prop max endurance
     B.   Both jet and prop max range
     C.   Prop max range, jet max endurance
     D.   Both jet and prop max endurance

4.   What is the effect of weight on maximum endurance and maximum range? What happens to maximum endurance and maximum range airspeed when weight is increased?

5.   What is the effect of altitude on maximum endurance and maximum range? What happens to maximum endurance and maximum range airspeed when altitude is increased?

6.   What effect does a tailwind have on maximum range and maximum endurance? Their airspeeds?

7.   What effect does lowering the landing gear have on fuel flow at maximum endurance airspeed?

8.   Define maximum angle of climb.

9.   How do jets and props achieve maximum angle of climb?
     A.   Full throttle, jets at L/D_{MAX}, props faster than L/D_{MAX}
     B.   85% throttle, jets at L/D_{MAX}, props slower than L/D_{MAX}
     C.   Full throttle, jets faster than L/D_{MAX}, props at L/D_{MAX}
     D.   Full throttle, jets at L/D_{MAX}, props slower than L/D_{MAX}

10.   Which statement is true concerning jet and prop climb performance?
        A.   Jets will always climb faster than props
        B.   Maximum rate of climb angle of attack is smaller than maximum angle of climb.
        C.   Maximum angle of climb airspeed is faster than maximum rate of climb.
        D.   Jets and props climb at the same angle of attack, but different airspeeds.

11.   What profile should a pilot fly to clear a tall obstacle on takeoff? What airspeed is this in the T-34C?

12.   How do altitude and weight changes affect maximum rate of climb and maximum angle of climb?

13.   What is the name for the altitude at which maximum power excess equals zero?

14.   At what AOA and velocity would you achieve maximum glide endurance?

15.   At what AOA and velocity would an airplane achieve maximum glide range?

16.   How do altitude, weight, and headwinds affect glide performance?

17.   From 10,000 feet AGL, approximately how far could a T-34C glide with no winds?
        A.   12 to 13 thousand feet
        B.   12 to 13 nautical miles
        C.   20 to 22 thousand feet
        D.   20 to 22 nautical miles

18.   What item of configuration will cause the greatest increase in sink rate?

19.   What are the throttle requirements in relation to airspeed in the region of reverse command?
        A.   Decrease in airspeed requires an increase in throttle
        B.   Decrease in airspeed requires a decrease in throttle
        C.   Increase in airspeed requires an increase in throttle
        D.   Constant airspeed requires an increase in throttle.

20.   In the region of reverse command, what is the effect of an increase in angle of attack with no change in throttle?

Aircraft Control Systems

INTRODUCTION

The purpose of this lesson is to aid the student in understanding the controls of the T-34C.

TERMINAL OBJECTIVE

Upon completion of this unit of instruction, the student aviator will demonstrate knowledge of basic aerodynamic factors that affect airplane performance.

ENABLING OBJECTIVES

1.116   State the type of control system used in the T-34C.

1.117   Describe how the control surfaces respond to control inputs.

1.118   Describe how the trim tab system holds an airplane in trimmed flight.

1.119   State the T-34C trim requirements for various conditions of flight.

1.120   State the point around which control surfaces are balanced.

1.121   Define mass balancing and aerodynamic balancing.

1.122   State the methods of mass balancing and aerodynamic balancing used by each control surface on the T-34C.

1.123   Describe how trim tabs can provide artificial feel.

1.124   State the purpose of bobweights and downsprings.

1.125   State the artificial feel devices used by each control surface on the T-34C.

Aircraft Control Systems

INTRODUCTION

This lesson introduces the control systems of the T-34C.

REFERENCES

1. Aerodynamics for Pilots
2. Introduction to the Aerodynamics of Flight
3. T-34C NATOPS Flight Manual

INFORMATION

To maneuver, the pilot must redirect the forces acting on the airplane. Control surfaces allow the pilot to change the amount of lift of the airfoil to which they are attached and create different airplane motions, such as yaw, pitch, and roll.

CONTROL SURFACES

The elevator is attached to the trailing edge of the horizontal stabilizer, and controls the pitching moment around the airplane's CG. Moving the control stick forward causes the elevator to move down, forcing the tail of the airplane up and pitching the nose down. Some airplanes move the entire horizontal stabilizer. This is called a stabilator and is used on the F-15 and F/A-18.

The ailerons are attached to the outboard trailing edges of the wings, and produce a rolling moment. Ailerons move in unison in opposite directions. If the control stick is pushed left, the left aileron rises, the right aileron lowers, and the plane rolls left. As long as the ailerons are deflected the airplane will continue to roll. When the stick is centered, the airplane will stop rolling, and remain at that bank angle until the stick is deflected again.

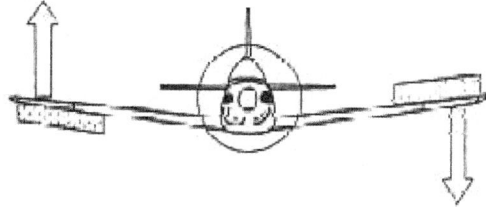

Figure 1-8-1 Aileron Operation

Spoilers may be attached to the wing's upper surface to provide roll control on some aircraft. Spoilers disrupt the airflow over the top of the wing in order to decrease the lift on the wing and cause the wing to roll downward. Spoilers may be used in conjunction with ailerons and/or stabilators.

The rudder is attached to the trailing edge of the vertical stabilizer, and produces a yawing moment. Stepping on the right rudder pedal moves the rudder to the right, pushing the tail left and yawing the airplane's nose to the right.

TRIM TABS

Trim tabs are attached to the trailing edge of each control surface and have two functions. The primary purpose of trim tabs is to trim. Trimming reduces the force required to hold control surfaces in a position necessary to maintain a desired flight attitude. Trim allows the pilot to fly virtually hands off, momentarily freeing the pilot's hands for other tasks, such as tuning radios or folding charts. The second purpose of trim tabs is to provide artificial feel (discussed later in this lesson).

If the pilot pulls back on the control stick, the elevator is deflected up so that the nose of the airplane pitches up. The airflow around the horizontal stabilizer creates a downward force on the elevator that acts at a distance (moment arm) from the hingeline. This creates a moment that tends to move the elevator back in line with the horizontal stabilizer. In order to keep the airplane's nose up, the pilot must exert enough back pressure on the control stick to overcome the moment created by the elevator's force. By moving the trim tab in the opposite direction as the control surface, a small force is created by the trim tab in the opposite direction. Since this small force has a greater moment arm, it creates a moment

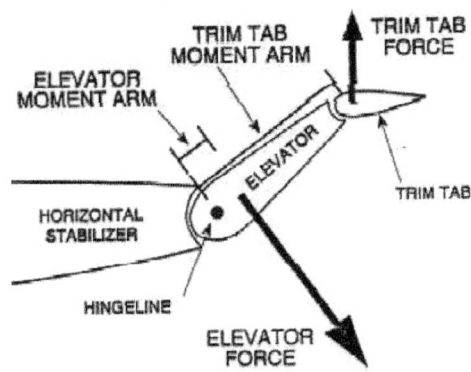

Figure 1-8-2 Trim Tab Operation

that exactly opposes the moment created on the elevator. Once the sum of the moments is zero, the elevator will remain in place until the pilot moves the control stick again. For trimming, trim tabs must always be moved in the opposite direction as the control surface. If a pilot moves a control away from its trimmed position, and then releases it, the trim tabs will cause the control surface to move back to its trimmed position. If the pilot moves the control surface and wants it to remain in place, he or she must re-trim it.

T-34C trim settings are changed by adjusting the manual trim wheels on the trim control panel. When the T-34C aileron trim tab wheel is adjusted, only the left aileron trim tab moves. The right aileron trim tab is preset by maintenance. T-34C rudder and elevator trim is adjusted frequently during flight because they are sensitive to power and airspeed changes. Rudder trim compensates for prop wash and torque, which vary with power. Elevator trim is adjusted to maintain various angles of attack while changing airspeed. Simply stated, right rudder trim is required for power increases and slower airspeeds,

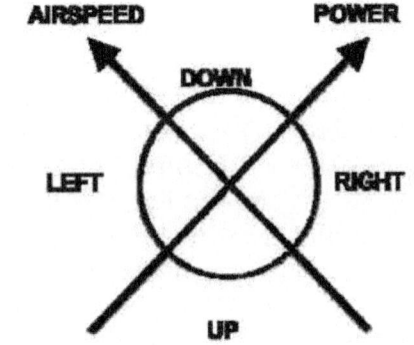

Figure 1-8-3 Rudder and Elevator Trim

while left rudder trim is required for power reductions and faster airspeeds (power changes take precedence at low speeds). Elevator trim is adjusted up at slower speeds and down at higher speeds (Figure 1-8-3).

CONTROL BALANCING

The forces that act at the control surface's center of gravity and aerodynamic center must be balanced around the hingeline in order to regulate control pressure, prevent control flutter, and provide control-free stability. Control-free is the situation where the controls are not being manipulated by the pilot (hands off). Aerodynamic balance concerns balancing the forces that act at the aerodynamic center. Mass balance concerns balancing the forces that act at the center of gravity.

Aerodynamic balance is used to keep control pressures (associated with higher velocities) within reasonable limits. As the trailing edge of the control surface is deflected in one direction, the leading edge deflects into the airstream forward of the hingeline (Figure 1-8-4). The force on the leading edge creates a moment that reduces the force required to deflect the control surface, so the pilot may control the airplane more easily. For aerodynamic balance, the T-34C uses shielded horns on the elevator and rudder, and an overhang on the ailerons (Figure 1-8-5).

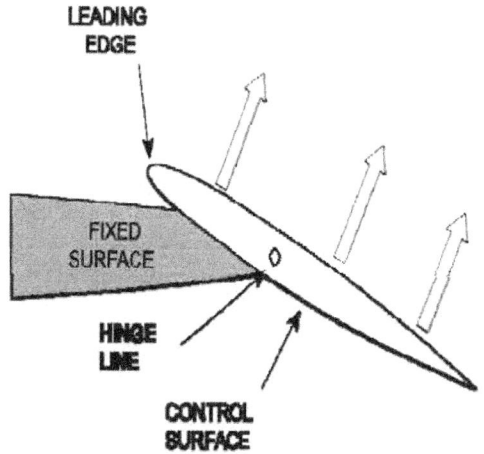

Figure 1-8-4 Aerodynamic Balance

The relationship of the control surface CG to the hingeline will determine the control-free stability of the airplane. Stability is more desirable in transport and bomber type airplanes and therefore the control surface CG is usually located forward of the hingeline. This keeps the control surface aligned with the fixed surface ahead of it when struck by gusts from turbulence. For high performance airplanes, the CG is located on or aft of the hingeline. With the CG aft of the hingeline, the control tends to float into the relative wind and cause a greater displacement which allows a faster response to control action and makes the airplane more maneuverable. To gain a balance between control response and stability, the T-34C control CGs are located on the hingeline. To locate the CG on the hingeline, weights are placed inside the control surface in the area forward of the hingeline (shielded horn and overhang). This technique is called **mass balancing**.

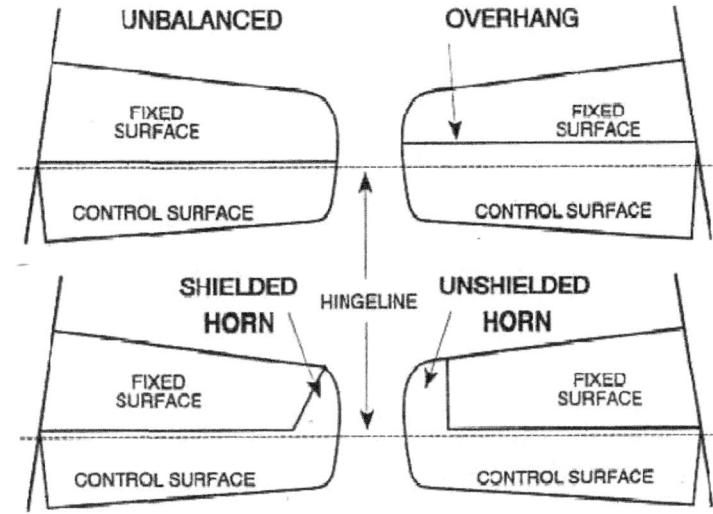

Figure 1-8-5 Aerodynamic Balance on a Horizontal Stabilizer

CONTROL FEEL

There are several basic types of control systems used to move the control surfaces: **conventional**, **power-boosted**, and **full power** (fly-by-wire).

In **conventional** controls the forces applied to the stick and rudder pedals are transferred directly to the control surfaces via push-pull tubes, pulleys, cables and levers. If an external force moves the control surfaces, the stick or rudder pedal will move in the cockpit. This action is called reversibility and gives the pilot feedback. Feedback is the force that the pilot feels in his hands or feet for a given deflection of the stick or rudder pedals. Without feedback the pilot would tend to over control and possibly overstress the airplane. The T-34C uses conventional controls.

Power-boosted controls have mechanical linkages with hydraulic, pneumatic, or electrical boosters to assist the pilot in moving the controls in the same way power steering assists a car driver. The degree to which the controls are boosted varies depending upon the mission and design of the airplane. These systems have some reversibility, and the pilot receives some control feel through the cockpit controls. If the boost system fails, the pilot can still control the airplane, although the control forces will be greatly increased.

With a **full-power** or fly-by-wire control system, the pilot has no direct connection with the control surfaces. The controls of a full power system are connected to hydraulic valves or electrical switches which control the movement of the control surfaces. The fly-by-wire system uses computer commands to displace the controls. These systems are not reversible. Movement of the control stick causes the control surfaces to move, but movement of the control surfaces will not cause the control stick to move. Since these systems are not reversible, they require an artificial means of producing control feel.

Artificial feel is the use any device used to create or enhance control feedback under various flight conditions such as airspeed and acceleration changes. The T-34C uses trim tabs, bobweights and down-springs to provide artificial feel to the pilot. The T-34C aileron uses a **servo trim tab** to provide artificial feel. It moves in the opposite direction as the aileron, thus helping the pilot deflect the aileron, making the airplane easier to maneuver.

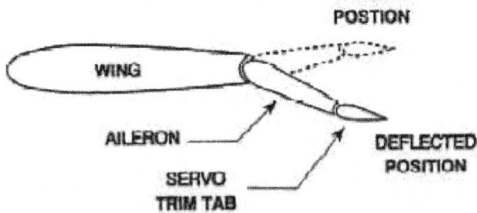

Figure 1-8-6 Servo Trim Tab

Artificial feel is provided to the T-34C rudder by an **anti-servo trim tab**. When the rudder is displaced, the anti-servo tab moves in the same direction at a faster rate. The more that a rudder pedal is pressed, the greater the resistance that the pilot will feel.

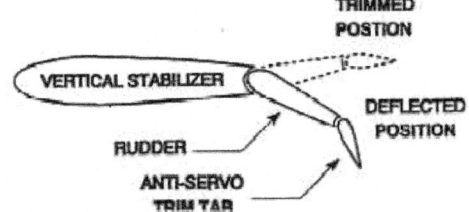

Figure 1-8-7 Anti-Servo Trim Tab

Because trim tabs do not provide the desired type of artificial feel, the T-34C elevator uses a **neutral trim tab** that maintains a constant angle to the elevator when the control surface is deflected. The elevator uses both a bobweight and a downspring to provide the pilot with some artificial feel. The downspring increases the force required to pull the stick aft at low airspeeds when required control pressures are extremely light. The bobweight increases the force required to pull the stick aft during maneuvering flight.

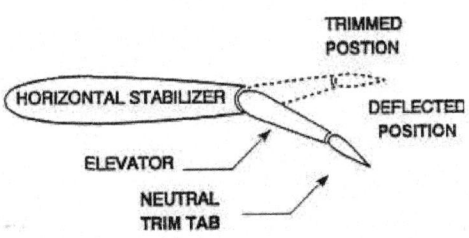

Figure 1-8-8 Neutral Trim Tab

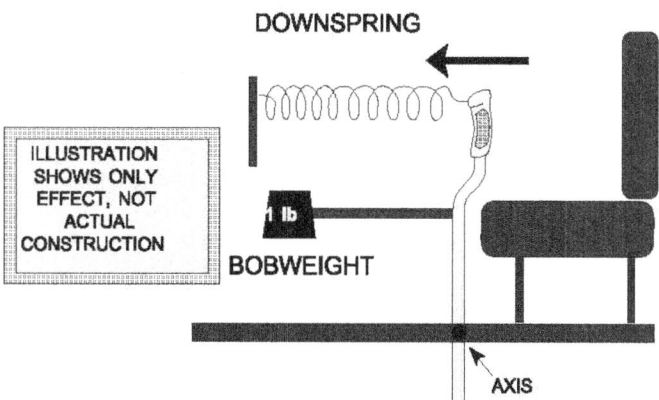

Figure 1-8-9 Elevator Artificial Feel

HISTORICAL NOTE

The problem of balancing stability and maneuverability has been around since flight began. It is interesting to note that the Wright brothers fully recognized this problem. They were the first to understand the need for positive roll control, such as ailerons. The Wright brothers were the first to demonstrate the use of ailerons with rudder for producing a coordinated turn (no side-slip). Interestingly they made their airplanes highly maneuverable by designing them to be statically unstable. Since then, most airplanes have been statically stable and relatively easy to fly.

1.   State the motion that each control surface creates.

2.   What control deflection is required to induce a nose-up pitching moment?

3.   What control deflection is required to induce a rolling moment to the right?

4.   What are the functions of the trim tabs? What elevator trim tab deflection is required to maintain equilibrium for a nose-up pitch attitude?

5.   What provides aerodynamic balance for the rudder and elevator of the T-34C?

6.   What is used for aerodynamic balancing of the T-34C aileron?

7.   What is responsible for mass balancing on the elevator of the T-34C?

8.   What type of controls does the T-34C use? How do cockpit control inputs move the control surfaces?

9.   What type of trim tabs are used on the T-34C aileron, elevator, and rudder? What type(s) assist the pilot?

10.   What is installed in the elevator system of the T-34C and is responsible for increasing the amount of force required to move the stick aft?

Stability

INTRODUCTION

The purpose of this lesson is to aid the student in understanding the stability of the T-34C.

TERMINAL OBJECTIVE

Upon completion of this unit of instruction, the student aviator will demonstrate knowledge of basic aerodynamic factors that affect airplane performance.

ENABLING OBJECTIVES

1.126   Define static stability and dynamic stability.

1.127   Identify the stability conditions of various systems based on their tendencies and motion.

1.128   Explain the relationship between stability and maneuverability.

1.129   State what may be done to increase an airplane's maneuverability.

1.130   Define longitudinal stability and neutral point.

1.131   Explain the contribution of straight wings, wing sweep, fuselage, horizontal stabilizer, and neutral point location to longitudinal static stability.

1.132   Define directional stability, sideslip angle, and sideslip relative wind.

1.133   Explain the contribution of straight wings, swept wings, fuselage, and vertical stabilizer to directional static stability.

1.134   Define lateral stability.

1.135   Explain the contribution of dihedral and anhedral wings, wing placement on the vertical axis, swept wings, and the vertical stabilizer to lateral static stability.

1.136   Describe directional divergence, spiral divergence, Dutch roll, and phugoid motion.

1.137   State the stability conditions that produce directional divergence, spiral divergence, Dutch roll, and phugoid motion.

1.138   Describe proverse roll, adverse yaw, and pilot induced oscillations.

1.139   Explain how pilot induced oscillations relate to the T-34C.

1.140   Describe the effects of asymmetric thrust, propeller slipstream swirl, P-factor, torque, and gyroscopic precession as they apply to the T-34C.

1.141   Describe what must be done to compensate for asymmetric thrust, propeller slipstream swirl, P-factor, torque, and gyroscopic precession as they apply to the T-34C

Stability

INTRODUCTION

The last few lessons have dealt mainly with weight, thrust and the production of lift and drag. All four of these forces are considered in equilibrium. During actual flight one or more of these forces may become unbalanced. Airplanes are designed to possess varying degrees of stability around the three rotational axes. This lesson will examine stability from an engineering standpoint.

REFERENCES

1.   Aerodynamics for Naval Aviators

2.   Aerodynamics for Pilots

3.   Introduction to the Aerodynamics of Flight

4.   T-34C NATOPS Flight Manual

INFORMATION

TYPES OF STABILITY

Stability is the tendency of an object (airplane) to return to its state of equilibrium once disturbed from it. There are two kinds of stability: static and dynamic. **Static stability** is the initial tendency of an object to move toward or away from its original equilibrium position. **Dynamic stability** is the position with respect to time, or motion of an object after a disturbance.

STATIC STABILITY

If an object has an initial tendency toward its original equilibrium position after a disturbance, it is said to possess **positive static stability**. Consider a ball inside a bowl (Figure 1-9-1). The ball's equilibrium position is at the bottom of the bowl. If the ball is moved from this position toward the rim of the bowl, its initial tendency, when released, is to roll back toward the bottom of the bowl. It is therefore said to possess positive static stability.

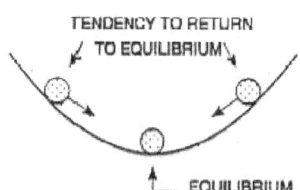

Figure 1-9-1 Positive Static Stability

Negative static stability is the initial tendency to continue moving away from equilibrium following a disturbance. Consider the bowl upside down with the ball on top as in Figure 1-9-2. Observe the ball's new equilibrium position. If the ball is moved away from its equilibrium position and released, its initial tendency is to roll farther away from equilibrium. The ball exhibits negative static stability.

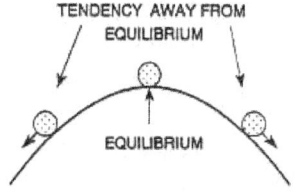

Figure 1-9-2 Negative Static Stability

Neutral static stability is the initial tendency to accept the displacement position as a new equilibrium. If we place the ball on a flat surface, it is again in equilibrium. If it is moved away from its original spot, the ball adopts the new equilibrium position (Figure 1-9-3). It does not have a tendency to move toward or away from the original equilibrium position. The ball now demonstrates neutral static stability.

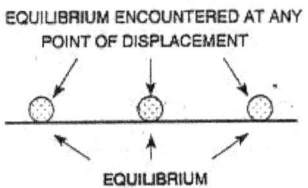

Figure 1-9-3 Neutral Static Stability

DYNAMIC STABILITY

Static stability reveals nothing about whether the object ever settles back to its original equilibrium position. To study dynamic stability, we will first assume the object to possess positive static stability.

Consider a ball at the top of Figure 1-9-1. After it is released, it will roll back to the bottom and up the other side. It will roll back and forth, oscillating less and less about the equilibrium position until it finally came to rest at the bottom of the bowl. It possesses **positive dynamic stability**. Note that although the ball passes through the equilibrium

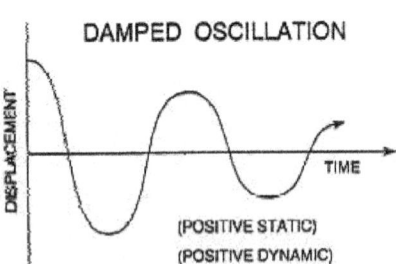

Figure 1-9-4 Positive Static and Positive Dynamic Stability

position, it is not in equilibrium again until it has stopped moving. The motion described is **damped oscillation** (Figure 1-9-4).

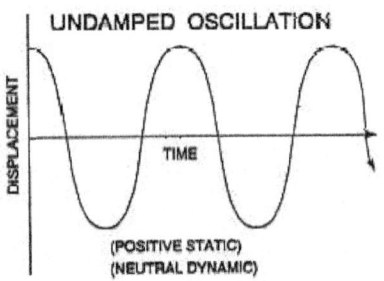

Figure 1-9-5 Positive Static and Neutral Dynamic Stability

If the ball oscillates about the equilibrium position and the oscillations never dampen out, it possesses **neutral dynamic stability**. Figure 1-9-5 depicts its displacement relative to equilibrium over time. This motion is **undamped oscillation**.

If, somehow, the ball did not slow down, but continued to climb to a higher and higher position with each oscillation, it would never return to its original equilibrium position. Figure 1-9-6 depicts **negative dynamic stability**. This motion is impossible with a ball, but occasionally aircraft behave this way. This motion is **divergent oscillation**.

If an object does not have positive static stability, it cannot have positive dynamic stability. If an object has positive static stability, it can have any dynamic stability. In other words, static stability does not ensure dynamic stability, but static instability ensures dynamic instability. If an object is dynamically **stable**, the displacement from equilibrium will be reduced until the object is again at its original equilibrium. It must have both positive static and positive dynamic stability. If an object is dynamically **unstable**, the displacement may or may not increase, but the object will never return to its original equilibrium.

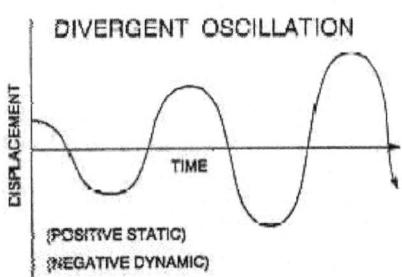

Figure 1-9-6 Positive Static and Negative Dynamic Stability

AIRPLANE STATIC STABILITY AND MANEUVERABILITY

Airplane stability is desirable because the airplane tends to stay in equilibrium. If an airplane were even slightly unstable, the pilot would have to spend most of the time keeping the airplane on the desired flight path. Most airplanes are built so that even if the pilot has no hands on the controls, they will continue to fly themselves.

Equilibrium occurs when the sum of the forces and moments around the center of gravity (CG) are equal to zero. An aircraft in equilibrium will travel in a constant direction at a constant speed, developing no moments that would cause it to rotate around the CG. Since an airplane can rotate around three different axes, we must consider its stability around each of these axes. **Longitudinal stability** is stability of the longitudinal axis around the lateral axis (pitch). **Lateral stability** is stability of the lateral axis around the longitudinal axis (roll). **Directional stability** is stability of the longitudinal axis around the vertical axis (yaw). Each motion requires a separate discussion.

We'll make some basic assumptions to simplify our study. First, we assume a constant TAS. The disturbances that cause the airplane to pitch, yaw, or roll will be small enough that it does not affect the airplane's forward velocity. The disturbances will also be small enough to keep the change in pitch attitude, and degree of yaw and roll small enough so that we do not approach any stalling AOAs or unusual attitudes. These disturbances are external and not caused by the pilot. The pilot applies no inputs to correct the displacement from equilibrium. Any moment that corrects the airplane's attitude results from the design of the airplane.

Any discussion of airplane stability requires an explanation on how the wings, fuselage, vertical stabilizer, horizontal stabilizer, etc, affect the longitudinal, lateral, and directional stability of the airplane. This lesson considers only conventional airplanes, that is, airplanes with their wings, fuselage and stabilizers in their normal positions.

An airplane's **maneuverability** is the ease with which it will move out of its equilibrium position. Maneuverability and stability are opposites. A stable airplane tends to stay in equilibrium and is difficult for the pilot to move out of equilibrium. The more maneuverable an airplane is, the easier it departs from equilibrium, and the less likely it is to return to equilibrium.

There are two ways to increase an airplane's maneuverability. If we want an airplane to move quickly from its trimmed equilibrium attitude, we can give it weak stability. Of course, this means the airplane will be more difficult to fly in equilibrium and will require more of the pilot's attention. Our other option is to give the airplane larger control surfaces. If the control surfaces are large, they can generate large moments by producing greater aerodynamic forces. The airplane designer must decide how to compromise between stability and maneuverability. The mission of a specific airplane dictates the compromises the designer will have to make. A transport plane must be relatively stable for long range flights and ease in landing. A fighter must possess great maneuverability for high performance turning.

LONGITUDINAL STATIC STABILITY

Now that we have a basic understanding of static stability, we can discuss each component and its individual contribution to static stability. Afterwards, we'll combine all the components and discuss the overall static stability of the airplane.

THE FLYING WING MODEL

Each individual component may have its own
aerodynamic center, and thus its own effect on static
stability. These individual components create
moments around the CG of our airplane that can be
either stabilizing or destabilizing. To examine stability
in greater detail, we will first take a simplified approach
using a "flying wing" model. By choosing the flying
wing we have essentially eliminated the stability effect
of any component except the wing itself. An airplane
experiences four main forces in equilibrium flight: lift,

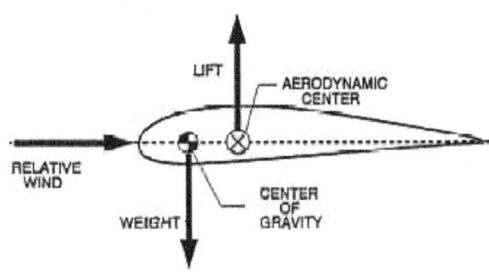

Figure 1-9-7 Flying Wing Model

weight, drag, and thrust. Recall that these forces act around the center of gravity. For our
discussion of longitudinal stability we only need to address lift and weight. Figure 1-9-7 shows
these two forces in equilibrium on our airplane.

Notice in Figure 1-9-7 that lift is acting through the aerodynamic center (AC), which is at a
distance from our CG. This creates a moment around the CG. It should be understood that
our flying wing in Figure 1-9-7 is in equilibrium, and that trim devices are preventing the wing
from rotating to a nose-down attitude.

Consider how the flying wing reacts to a disturbance that increases the AOA sensed by the
airfoil. The increased AOA will increase lift. If the CG is ahead of the AC, the increase in lift at
the AC develops a moment that pitches the nose of the airplane down in the direction of its
original equilibrium AOA. Our flying wing has positive longitudinal static stability because of its
initial tendency to return to equilibrium (Figure 1-9-8). If a component's aerodynamic center is
behind the airplane's center of gravity the component will be a positive contributor to
longitudinal static stability.

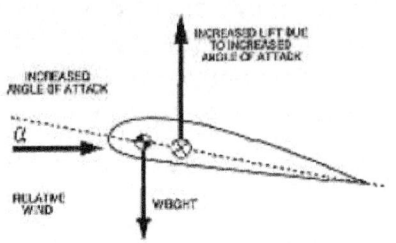

Figure 1-9-8 Positive Longitudinal
Static Stability

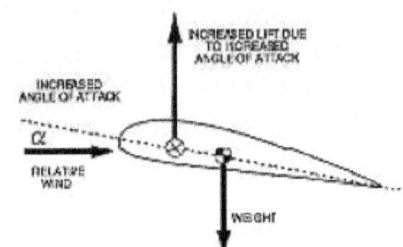

Figure 1-9-9 Negaitve Longitudinal
Static Stability

Next, examine a flying wing where the AC is ahead of the CG. When the disturbance
increases AOA, the wing produces more lift and rotates the flying wing further away from
equilibrium (Figure 1-9-9). Any disturbance would soon lead to stall and possibly out of control
flight. We can generalize this and say that if a component's aerodynamic center is in front of
the airplane's center of gravity the component will be a negative contributor to longitudinal
static stability.

STRAIGHT WINGS

The wing's contribution to longitudinal static stability depends mainly on the location of the wing's AC with respect to the airplane's CG. Most airplanes have straight wings with the AC forward of the airplane's CG. Like the second flying wing example, having the AC forward of the CG causes longitudinal static instability. The wings of most conventional airplanes are negative contributors to longitudinal static stability. Figure 1-9-10 illustrates the location of the wing's AC and the airplane's CG.

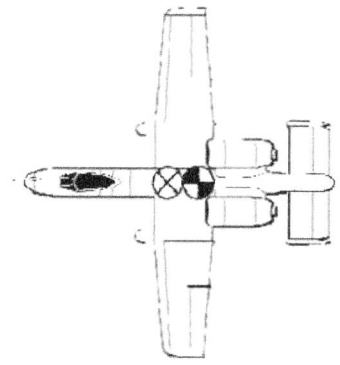

Figure 1-9-10 Straight Wings

WING SWEEP

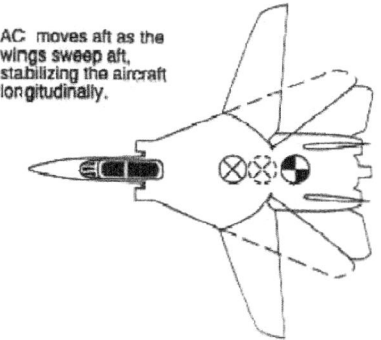

AC moves aft as the wings sweep aft, stabilizing the aircraft longitudinally.

When an F-14's wings are swept forward, they have a strong longitudinally destabilizing effect. This is because the wing's AC is well forward of the airplane's CG. This increases the F-14's maneuverability. As sweep angle increases (i.e. the wings move aft), the wings' AC moves aft, closer to the airplane's CG (Figure 1-9-11), making the airplane more longitudinally stable. Sweeping an airplane's wings back is a positive contributor to longitudinal static stability.

THE FUSELAGE

Figure 1-9-11 Wing Sweep

The fuselage acts as an airfoil and thus produces lift. The fuselage's AC is usually located ahead of the airplane's CG (Figure 1-9-12). If a disturbance causes an increase in angle of attack, the fuselage will produce greater lift that produces a destabilizing effect. The fuselage is a negative contributor to longitudinal stability.

THE HORIZONTAL STABILIZER

The horizontal stabilizer is a symmetric airfoil designed to stabilize the airplane around the lateral axis. Its contribution to longitudinal static stability is determined by the moment it produces around the CG. Since its AC is well behind the airplane's CG (Figure 1-9-13), the horizontal stabilizer has the greatest positive effect on longitudinal static stability. The pitching moment can be increased by increasing the distance between the airplane's CG and the stabilizer's AC, or by enlarging the horizontal stabilizer. Thus, for a short airplane, a large horizontal stabilizer is needed. For an airplane with a longer moment arm, a smaller horizontal stabilizer will suffice.

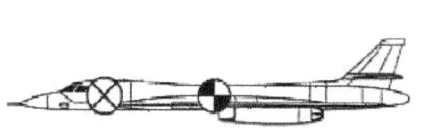

Figure 1-9-12 Fuselage

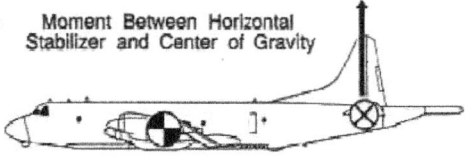

Moment Between Horizontal Stabilizer and Center of Gravity

Figure 1-9-13 Horizontal Stabilizer

THE NEUTRAL POINT

The longitudinal static stability provided by the horizontal stabilizer must overcome the instabilities of the wings and fuselage in order to produce a stable airplane. Figure 1-9-14 shows the AC for each individual component. The **neutral point (NP)** is the location of the center of gravity along the longitudinal axis that would provide neutral longitudinal static stability. It can be thought of as the aerodynamic center for the entire airplane. The location of the NP is fixed on conventional airplanes, but we can change

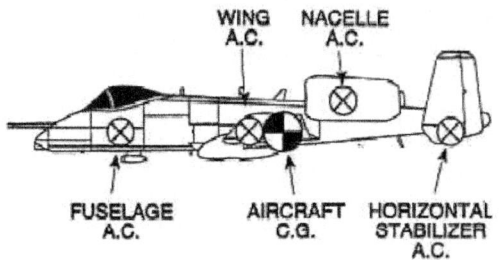

Figure 1-9-14 AC for each component

the location of the CG by moving around cargo or mounting ordnance and fuel in various locations. As the CG is moved aft, the airplane's static stability decreases. The NP defines the farthest aft CG position without negative stability. Once the CG is aft of the NP the airplane becomes unstable and the pilot may have difficulty controlling it in flight. The neutral point does not determine the contribution of individual components, but rather determines the longitudinal stability of the overall airplane.

DIRECTIONAL STATIC STABILITY

Directional static stability is stability of the longitudinal axis around the vertical axis. When an airplane yaws, its momentum keeps it moving along its original flight path for a short time. This condition is known as a **sideslip**. The angle between the longitudinal axis and the relative wind is called the **sideslip angle (β)**. The component of the relative wind that is parallel to the lateral axis is called the **sideslip relative wind**. Reaction to the sideslip will determine a component's contribution to directional static stability. We will examine the effects of the wings, wing sweep, fuselage, and the vertical stabilizer on directional static stability.

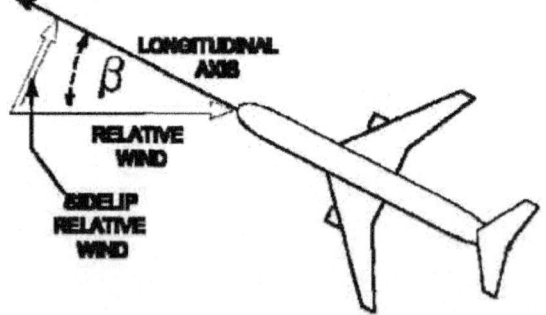

Figure 1-9-15 Sideslip Angle β

STRAIGHT WINGS

During a sideslip, the advancing wing on a straight winged airplane has a momentary increase in airflow velocity as it moves forward. This increases parasite drag on that wing and pulls it back to its equilibrium position. The retreating wing has less velocity and less parasite drag, which helps to bring the nose back into the relative wind. Therefore, straight wings have a small positive effect on directional static stability.

SWEPT WINGS

The swept design of a wing will further increase directional stability. The advancing wing not only experiences an increase in parasite drag, but also an increase in induced drag due to the increased chordwise flow. Remember that lift and induced drag are produced by the wings when air flows chordwise over them. The retreating wing experiences more spanwise flow. Figure 1-9-16 depicts this phenomenon with the left wing experiencing greater induced and parasite drag. The result is an airplane that comes back into the relative wind.

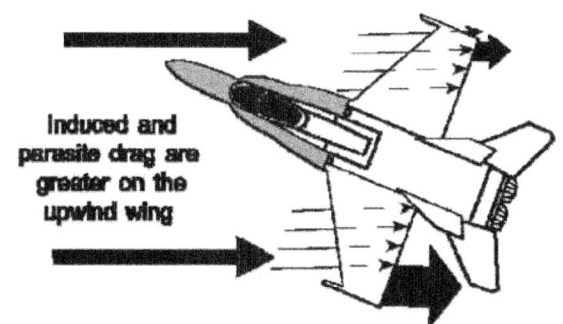

Figure 1-9-16 Swept Wings

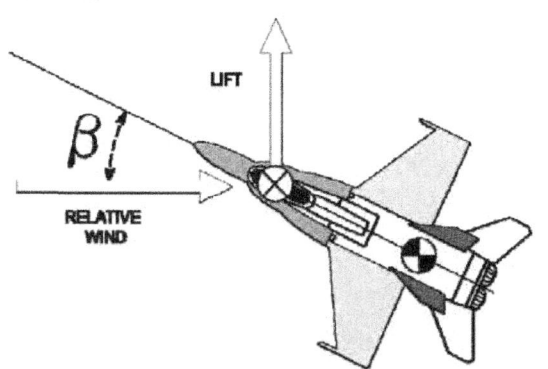

Figure 1-9-17 Fuselage

THE FUSELAGE

The fuselage is a symmetric airfoil with its aerodynamic center forward of the airplane's CG. At zero angle of attack or zero sideslip it produces no net lift. When the airplane enters a sideslip, an angle of attack is created on the fuselage. The lift produced at the fuselage AC pulls the nose away from the relative wind, thus causing an increase in the sideslip angle. Therefore, the fuselage is a negative contributor to the airplane's directional static stability (Figure 1-9-17).

THE VERTICAL STABILIZER

The vertical stabilizer is the greatest positive contributor to the directional static stability of a conventionally designed airplane. The vertical stabilizer is a symmetric airfoil mounted far behind the airplane's CG. A sideslip causes the vertical stabilizer to experience an increased angle of attack. This creates a horizontal lifting force on the stabilizer that is multiplied by the moment arm distance to the airplane's CG (Figure 1-9-18). The moment created will swing the nose of the airplane back into the relative wind. This is identical to the way a weathervane stays oriented into the wind. There is an

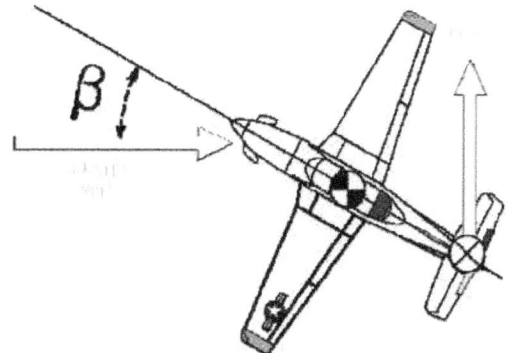

Figure 1-9-18 Vertical Stabilizer

inverse relationship between tail size and moment arm length. The smaller the distance to the CG, the larger the vertical stabilizer must be and vice versa. It is not always desirable to have a large vertical stabilizer because of limited storage room aboard aircraft carriers and the large radar signatures. Designers often use two or more smaller vertical stabilizers (A-10, F-15, F/A-18, and E-2), to accomplish the same stability effects as one large stabilizer.

LATERAL STATIC STABILITY

Lateral stability is stability of the lateral axis around the longitudinal axis. An airplane has lateral stability if, after some disturbance causes it to roll, it generates forces and moments that tend to reduce the bank angle and restore the airplane to a wings level flight condition. When an airplane rolls, the lift vector points to the inside of the turn, reducing the vertical component of lift. Since weight still acts downward with the same force (Figure 1-9-19), the plane descends. The horizontal component of lift pulls the airplane to the side, thus creating a sideslip relative wind. This sideslip relative wind acts on the various components of the airplane causing stability or instability.

DIHEDRAL EFFECT

When an airplane is laterally sideslipping, dihedral wings cause an increase in angle of attack and lift on the down-going wing. The up-going wing has a reduced angle of attack and a decrease in lift (Figure 1-9-20). This difference in lift creates a rolling moment that rights the airplane and stops the sideslip. Wings that are straight have neutral lateral static stability. Dihedral wings are the greatest positive contributors to lateral static stability. Anhedral wings are the greatest negative contributors to lateral static stability.

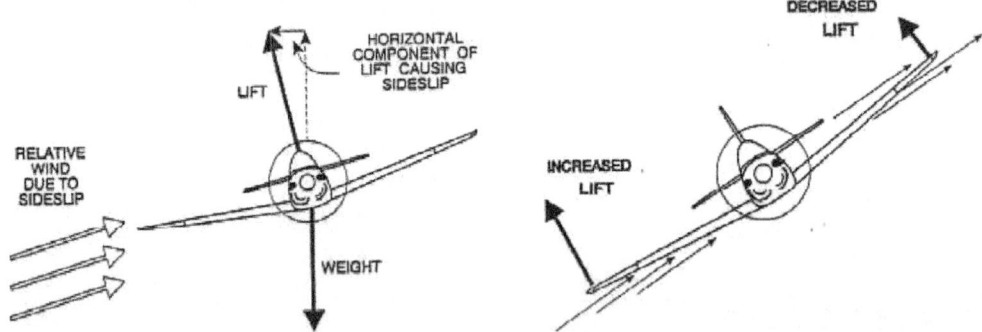

Figure 1-9-19 Sideslip Relative Wind   Figure 1-9-20 Dihedral Wings

WING PLACEMENT ON THE VERTICAL AXIS

During a sideslip the fuselage of a high-winged airplane impedes the airflow generated by the sideslip. This increases the upwash at the wing root on the down-going wing which increases the AOA and lift. The up-going wing receives downwash which decreases the AOA, and lift. The lift imbalance rolls the airplane back to wings level. A low-mounted wing has the opposite effect. Thus, a high mounted wing is a positive contributor, and a low mounted wing is a negative contributor to lateral static stability.

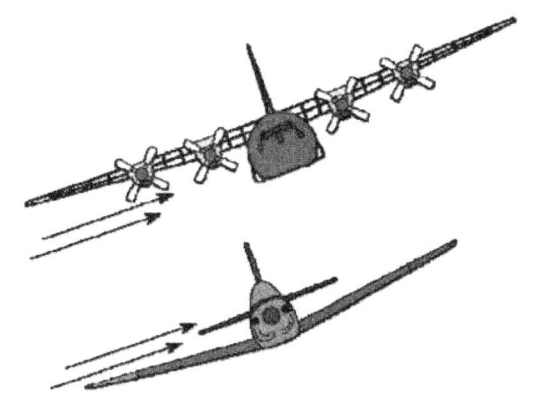

Figure 1-9-21 High- and Low-Mounted Wings

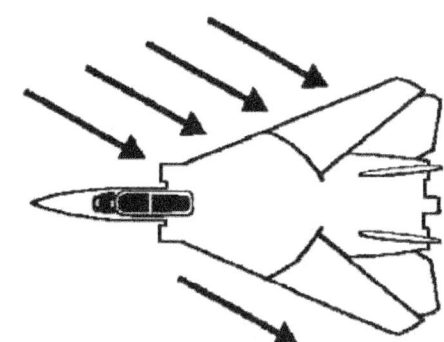

Figure 1-9-22 Wing Sweep

WING SWEEP

Another way to affect lateral stability is to sweep the wings aft. As an airplane begins to sideslip in the direction of roll, the wing toward the sideslip has more chordwise flow than the wing away from the sideslip (Figure 1-9-22). The wing toward the sideslip (the lower wing) generates more lift, which levels the wings. Swept wings are laterally stabilizing. These effects are cumulative. High-mounted, swept dihedral wings are much more stable than low-mounted, straight wings with the same dihedral.

THE VERTICAL STABILIZER

The only other major effect on lateral stability comes from the vertical stabilizer. When in a lateral sideslip, the vertical stabilizer senses an angle of attack, so it produces lift. Since the tail is above the airplane's center of gravity, this lift produces a moment that tends to right the airplane (Figure 1-9-23).

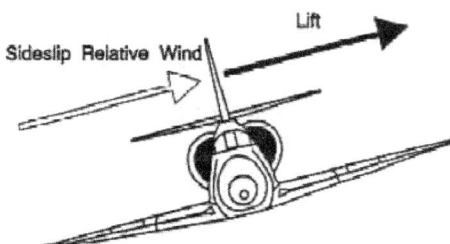

Figure 1-9-23 Vertical Stabilizer

DYNAMIC STABILITY

Our discussion thus far has centered on static stability. When we discuss dynamic stability, we must realize that lateral and directional stability are interrelated. This interrelationship is called cross-coupling. The motions of an airplane are such that a rolling motion causes a yawing motion and vice versa. This cross-coupling between directional static stability and lateral static stability causes several dynamic effects including spiral divergence, Dutch roll, proverse roll, and adverse yaw.

DIRECTIONAL DIVERGENCE

Directional divergence is a condition of flight in which the reaction to a small initial sideslip results in an increase in sideslip angle (Figure 1-9-24). Directional divergence is caused by negative directional static stability. If the vertical stabilizer becomes ineffective for some reason (battle damage, mid-air collision), directional divergence could cause out of control flight. Most airplanes have very strong directional stability to prevent this from occurring.

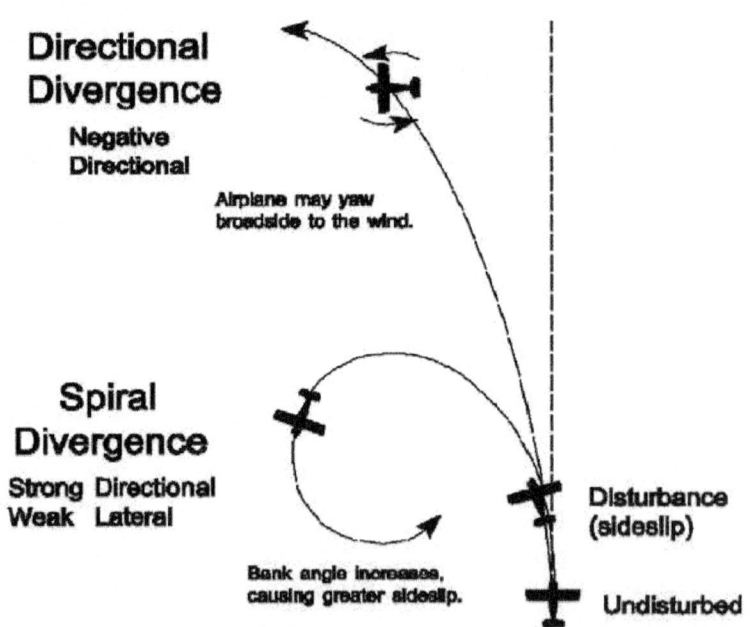

Figure 1-9-24 Directional and Spiral Divergence

SPIRAL DIVERGENCE

Spiral divergence occurs when an airplane has strong directional stability and weak lateral stability (Figure 1-9-24). For example, an airplane is disturbed so that its wing dips and starts to roll to the left. Because the airplane has weak lateral stability it cannot correct itself and the flight path arcs to the left. The airplane senses a new relative wind from the left and aligns itself with the new wind by yawing into it (strong directional stability). The right wing is now advancing and the increased airflow causes the airplane to roll even more to the left. The airplane will continue to chase the relative wind and will develop a tight descending spiral. This is easily corrected by control input from the pilot.

DUTCH ROLL

Dutch roll is the result of strong lateral stability and weak directional stability. The airplane responds to a disturbance with both roll and yaw motions that affect each other. For example, a gust causes the airplane to roll left, producing a left sideslip. The strong lateral stability increases lift on the left wing and corrects it back to wings level. At the same time, the nose of the airplane yaws left into the sideslip relative wind. This leaves the airplane wings level, with the nose cocked out to the left.

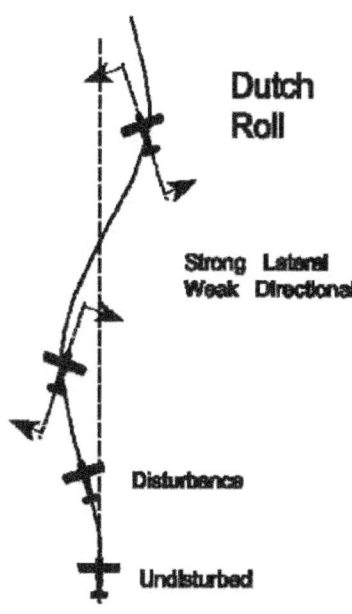

Figure 1-9-25 Dutch Roll

The weak directional stability now swings the nose to the right to correct the nose back into the relative wind. This causes the left wing to advance faster than the right wing, a situation which produces more lift on the left wing and rolls the airplane to the right, creating a right sideslip. The strong lateral stability corrects the wings back to level, while the nose yaws right into the sideslip relative wind. This leaves the airplane wings level, with the nose cocked out to the right. As the nose yaws left into the relative wind, the wings will roll left which starts the entire process again.

The airplane appears to be "tail wagging" (Figure 1-9-25). This condition can be tolerated and may eventually dampen out. However, it is not acceptable in a fighter or attack airplane when the pilot is trying to aim at a target.

PHUGOID OSCILLATIONS

Phugoid oscillations are long period oscillations (20 to 100 seconds) of altitude and airspeed while maintaining a nearly constant angle of attack. Oscillations of pitch attitude do occur, but are often minor. Upon being struck by an upward gust, an airplane would gain altitude and lose airspeed. A large but gradual change in altitude and airspeed occurs. When enough airspeed is lost, the airplane will nose-over slightly, commencing a gradual descent, gaining airspeed and losing altitude. When enough airspeed is regained, the nose will pitch up, starting the process over. The period of this oscillation is long enough that the pilot can easily correct it. Often, due to the almost negligible changes in pitch, the pilot may make the necessary corrections while being completely unaware of the oscillation.

PILOT / AIRPLANE INTERACTION

A complete discussion of an airplane's stability characteristics is not limited to how the airplane reacts to various external forces, but must also consider the interaction of the pilot and the airplane.

PROVERSE ROLL

Proverse roll is the tendency of an airplane to roll in the same direction as it is yawing. When an airplane yaws, the yawing motion causes one wing to advance and the other wing to retreat. This increases the airflow on the advancing wing and decreases airflow over the retreating wing. A difference in lift is created between the two wings, and the airplane rolls in the same direction as it yawed. Proverse roll is even more pronounced on swept wing airplanes since the advancing wing will have more chordwise flow and will produce more lift.

ADVERSE YAW

Adverse yaw is the tendency of an airplane to yaw away from the direction of aileron roll input. When an airplane rolls, it has more lift on the up-going wing than on the down-going wing. This causes an increase in induced drag on the up-going wing that will retard that wing's forward motion and cause the nose to yaw in the opposite direction of the roll. The aircraft produces adverse yaw each time the ailerons are deflected (rolling into and out of a turn).

We can do three things to overcome this problem. The first method is to use spoilers instead of ailerons. The spoiler is deflected into the airstream from the upper surface of the wing. This spoils the airflow and reduces lift, causing the airplane to roll. The spoiler increases the para- site drag on the down-going wing, offsetting the induced drag on the up-going wing and helps reduce or eliminate adverse yaw. The second method is to use a rudder input to offset ad- verse yaw. The third is actually a design method of building the aircraft with differential ailerons.

PILOT INDUCED OSCILLATIONS

Pilot induced oscillations (PIO) are short period oscillations of pitch attitude and angle of attack. PIO or porpoising occurs when a pilot is trying to control airplane oscillations that happen over approximately the same time span as it takes to react. For example, a gust of wind causes the nose to pitch up. The natural longitudinal stability of the airplane will normally compensate. However, if the pilot tries to push the nose-down, his input may coincide with the stability cor- rection, causing the nose to over correct and end up low. The pilot then pulls back on the stick causing the nose to be high again. Since the short period motion of PIO is of relatively high frequency, the amplitude of the pitching could reach dangerous levels in a very short time. If PIO is encountered, the pilot must rely on the inherent stability of the airplane and immediately release the controls, if altitude permits. If not, the pilot should "freeze" the stick slightly aft of neutral. The T-34C is not subject to this type of oscillation since it does not have strong longitudinal static stability.

ASYMMETRIC THRUST

Any airplane with more than one engine can have directional control problems if one engine fails. This is known as asymmetric thrust. If an airplane with its engines located far from the fuselage, such as an S-3, E-2 or KC-10, has an engine failure, the thrust from the operating engine(s) will create a yawing moment toward the dead engine. This can happen even if the engines are relatively close, such as with the F/A-18. The farther from the longitudinal axis that the engines are located, the greater the moment created by the operating engine. The yawing motion may be sufficient to cause proverse roll. Full opposite rudder may be required to compensate for the yawing moment, while opposite aileron should be used to correct the proverse roll. Every multi-engine airplane has a minimum directional control speed that must be flown to ensure maximum effectiveness of the vertical stabilizer following an engine failure.

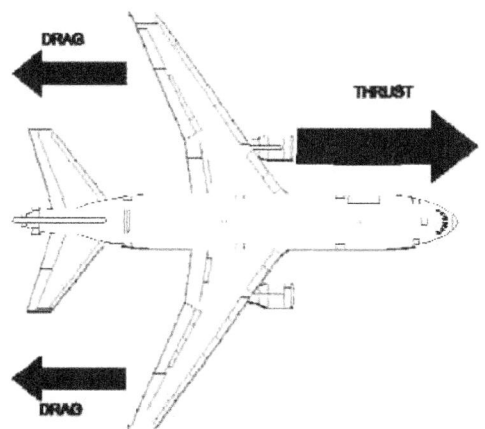

Figure 1-9-26 Asymmetric Thrust

SLIPSTREAM SWIRL

The propeller imparts a corkscrewing motion to the air called the slipstream swirl. This corkscrewing air flows around the fuselage until it reaches the vertical stabilizer where it increases the AOA on the vertical stabilizer (Figure 1-9-27). When a propeller driven airplane is at a high power setting and low airspeed (e.g., during takeoff), the increased angle of attack creates a horizontal lifting force that pulls the tail to the right and causes the nose to yaw left. Right rudder and lateral control stick inputs are required to compensate for slipstream swirl.

Figure 1-9-27 Propeller Slipstream Swirl

P-FACTOR

Propeller factor (P-factor) is the yawing moment caused by one prop blade creating more thrust than the other. The angle at which each blade strikes the relative wind will be different (Figure 1-9-28), causing a different amount of thrust to be produced by each blade. For practical purposes, only the up-going and down-going blades are considered. If the relative wind is above the thrust line, the up-going propeller blade on the left side creates more thrust since it has a larger angle of attack with the relative wind. This will yaw the nose to the right (Figure 1-9-29). Note that this right yaw will result at high airspeeds (above 150 knots for the T-34C) due to the slight nose-down attitude required in level flight. If the relative wind is below the thrust line, such as in flight near the stall speed, the down-going blade on the right side will create more thrust and will yaw the nose to the left.

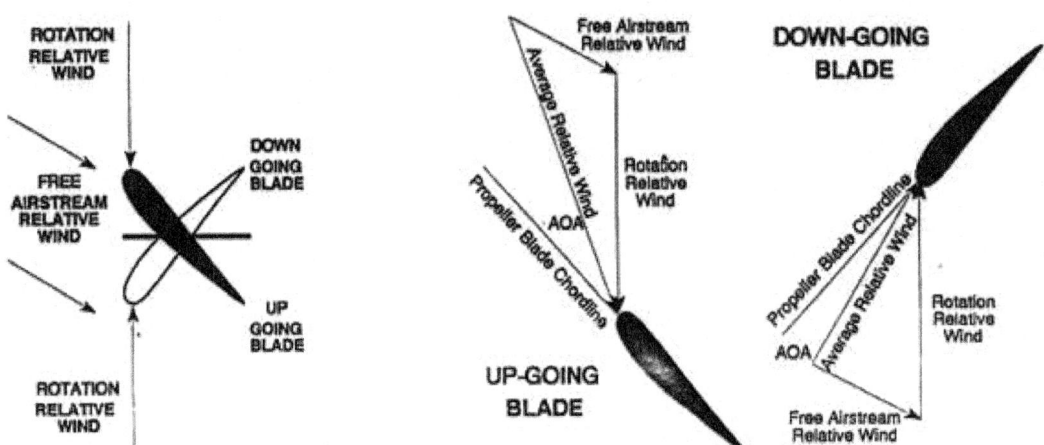

Figure 1-9-28 Propeller Side View   Figure 1-9-29 Relative Wind (Nose-Down Attitude)

There are two basic requirements for P-factor to have a noticeable effect: The engine must be set to a high power condition, and the thrust axis must be displaced from the relative wind. Since airplane designers want P-factor to be minimized during the majority of flight, they align the thrust axis with the relative wind for cruise airspeeds. Thus, P-factor will be most prevalent at AOAs significantly different from cruise AOA, such as very high speed level or descending flight, and high angle of attack climbs.

TORQUE

Torque is a reactive force based on Newton's Third Law of Motion. A force must be applied to the propeller to cause it to rotate clockwise. A force of equal magnitude, but opposite direction, is produced which tends to roll the airplane's fuselage counter-clockwise. The T-34C uses the elevator trim tabs to compensate for torque. If the elevator trim is set to zero, the left trim tab is approximately 4.5° down from the elevator, while the right trim tab is approximately 4.5° up from the elevator.

A turbojet aircraft will not experience torque from its engines. Jet engines do not push against the airframe in order to rotate, they rest on bearings and push against the airflow to rotate. The torque in a turboprop is applied through its gearbox, not its engine.

GYROSCOPIC PRECESSION

Gyroscopic precession is consequence of the properties of spinning objects. When a force is applied to the rim of a spinning object (such as a propeller) parallel to the axis of rotation, a resultant force is created in the direction of the applied force, but occurs 90° ahead in the direction of rotation. Pitching the nose of the T-34C down produces an applied force acting forward on the top of the propeller disk. The resultant force would act 90° ahead in the direction of propeller rotation (clockwise), and cause the T-34C to yaw left. Gyroscopic precession often plays a large role in determining an airplane's entry characteristics into a spin.

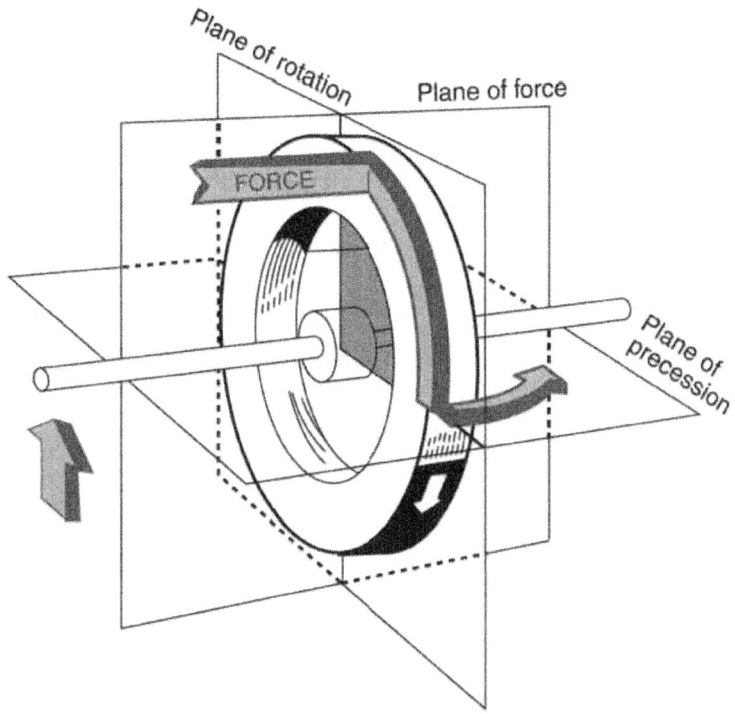

Figure 1-9-30 Gyroscopic Precession

1. Define static and dynamic stability.

2. What is the characteristic stability of divergent oscillation?

3. What is the relationship between stability and maneuverability?

4. What motion does longitudinal stability concern?

5. How do the fuselage, horizontal stabilizer, wing, wing sweep, and location of the CG affect the longitudinal stability of an airplane?

6. What effect does increasing wing sweep angle have on the location of the wing's aerodynamic center?

7. What factors determine the magnitude of the contribution of the horizontal stabilizer to longitudinal stability?

8. Define sideslip angle.

9. What motion does directional stability concern?

10. How do the fuselage, straight wing, swept wing, and vertical stabilizer affect directional stability of an airplane?

11.   What motion does lateral stability concern?

12.   How do swept wings, dihedral, anhedral, high or low mounted wings, and the vertical stabilizer affect lateral static stability?

13.   What types of stability are associated with directional divergence, spiral divergence, and Dutch roll?

14.   What is the tendency of an airplane to yaw away from the direction of roll? What causes it?

15.   What is the tendency of an airplane to roll into the direction of a yaw? What causes it?

16.   What must be done to achieve zero sideslip in case of an asymmetric thrust condition?

17.   If the relative wind is below the thrust line, which blade will produce more thrust, the down-going or up-going blade? Why? Which way will the nose of the airplane yaw?

18.   If the PCL is suddenly pushed forward, the T-34C will tend to roll to the _____ due to an increase in _____. If the nose of the airplane is suddenly pushed down, gyroscopic precession will tend to yaw the T-34C to the _____.

Spins

INTRODUCTION

The purpose of this lesson is to aid the student in understanding the aerodynamic characteristics of spins and spin recovery.

TERMINAL OBJECTIVE

Upon completion of this unit of instruction, the student aviator will demonstrate knowledge of basic aerodynamic factors that affect airplane performance.

ENABLING OBJECTIVES

1.142	Define spin and autorotation.
1.143	Identify the factors that cause a spin.
1.144	Identify the effects of weight, pitch attitude, and gyroscopic effect on spin entry.
1.145	Describe the angles of attack and forces on each wing that cause autorotation during a spin.
1.146	State the characteristics and cockpit indications of normal and inverted spins.
1.147	Identify the effects of control inputs on spin recovery.
1.148	State how the configuration of the empennage and placement of the horizontal control surfaces can affect spin recovery.
1.149	Describe the spin recovery procedures for the T-34C.
1.150	Define progressive and aggravated spin.

Spins

INTRODUCTION

In the early days of aviation, a spin usually ended in a fatality. Aerodynamic theory had not advanced to the point of being able to explain a spin or determine the proper recovery techniques. Even today, the word "spin" has an aura of danger and the unknown. However, with an understanding of spins, your apprehension will be measurably reduced.

Once spins were understood, nearly all pilots have practiced recovering from them. The spin itself, however, has no practical value as a maneuver. There are at least three sound reasons for teaching spins to student pilots. First, virtually every aircraft that is capable of stalling has the potential for entering a spin. In most high performance aircraft, many maneuvers are flown near the stall region, making it essential to train aircrews to recognize the conditions leading to spins, apply appropriate spin recovery procedures, and learn to respect spins without fearing them. Second, spin training builds confidence in one's ability to handle an aircraft should it inadvertently enter a spin. Third, spin training improves a pilot's ability to remain oriented and still make appropriate control inputs.

Every aircraft has different spin characteristics and recovery techniques. Therefore, the pilot must know the flight manual procedures for spin prevention and recovery for the aircraft he is flying. This lesson will cover general spin characteristics and specific indications of a spin along with spin prevention/recovery steps for the T-34C.

REFERENCES

1.   Aerodynamics for Naval Aviators

2.   Aerodynamics for Pilots

3.   T-34C NATOPS Flight Manual

INFORMATION

SPIN AERODYNAMICS

A **spin** is an aggravated stall that results in autorotation. **Autorotation** is a combination of roll and yaw that propagates itself and progressively gets worse due to asymmetrically stalled wings. For an aircraft to spin, it must be stalled and some form of yaw must be introduced. If an aircraft is not stalled, it will not spin. If either the stall or yaw is removed, the aircraft will not continue to spin. A yawing moment can be induced with the rudder, by adverse yaw, gust wing loading, etc. If a stalled condition is maintained long enough, sufficient yaw to enter a spin could eventually be introduced.

To help you understand the aerodynamics of the spin, consider the motions an aircraft undergoes during a spin. Every aircraft exhibits different spin characteristics, but they all have stall and yaw about the spin axis. For the T-34C, the spin axis is through the cockpit.

Examine the AOA and relative wind on each wing (Figure 1-10-1). Suppose the airplane stalls and begins a roll to the left. The left wing becomes the down-going wing and senses a roll relative wind from beneath. This roll relative wind is added to the existing relative wind and creates an average relative wind that is further from the chordline. Therefore, the down-going wing has a higher AOA. This wing is more stalled. Conversely, the right wing becomes the up-going wing and senses a roll relative wind from above. When added to the original relative wind, the up-going wing has a lower AOA and is less stalled. Remember that while both wings are stalled, they do not lose all their lift, nor are they equally stalled.

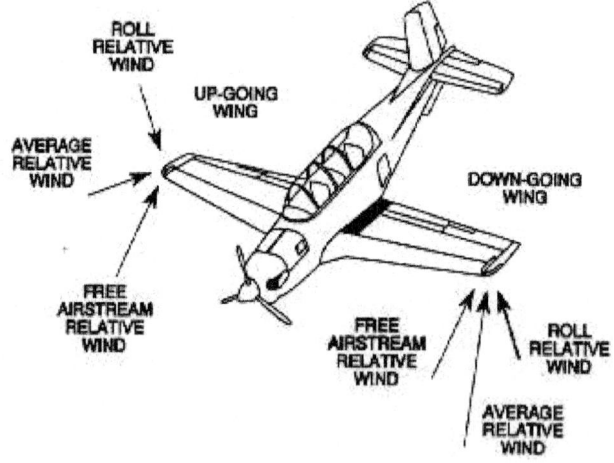

Figure 1-10-1 Roll Relative Wind

The increased AOA on the down-going wing decreases the C_L generated by that wing. The up-going wing has a greater C_L due to its smaller AOA, and therefore has greater total lift (Figure 1-10-2). The greater total lift on the up-going wing results in a continued rolling motion of the airplane. The down-going wing has a higher C_D due to its increased AOA. The greater drag on the down-going wing results in a continued yawing motion in the direction of roll. The combined effects of roll and yaw cause the airplane to continue its autorotation.

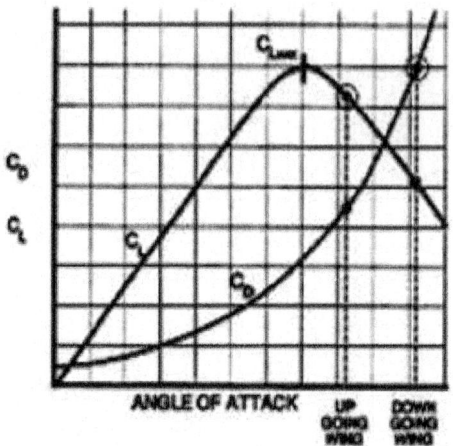

Figure 1-10-2 C_L and C_D in a Spin

SPIN INDICATIONS

The T-34C will spin either erect (upright) or inverted. Erect spins result from positive-g stall entries. Inverted spins can occur from either a negative-g stall or an improperly applied recovery from an erect spin that results in a negative-g stall. The type of spin is not dependent upon aircraft attitude at the time of the stall.

In case of spatial disorientation during a spin, the pilot must be aware of what the cockpit instrument spin indications are for each type of spin. The instruments used to confirm an actual spin are the turn needle, AOA indicator, and airspeed indicator. The turn needle is the only reliable indicator of spin direction. The balance ball (slip indicator) gives no useful indication of spin direction and should be disregarded. The altimeter is monitored to ensure compliance with bailout/ejection criteria.

Gauge	Spin Indications	Remarks
Altimeter	Rapidly decreasing	May indicate up to 1000 feet above actual altitude.
AOA	30 units (pegged)	Stalled
Airspeed	80–100 knots	Stable
Turn Needle	Pegged in direction of spin	Spin rate: 100–170° per sec.
VSI	6000 fpm (pegged)	9,000–12,000 fpm
Attitude Gyro	May be tumbling	45° Nose down

Table 1-10-1 T-34C Indications of an Erect Spin
(characterized by nose-down, upright attitude, and positive Gs)

An inverted spin is characterized by an inverted attitude and negative Gs on the airplane. Stabilized inverted spins are uncommon because the positioning of the vertical stabilizer in this spin causes the airplane to recover easily. Inverted spins are very disorienting to the aircrew and difficult to enter. The T-34C is prohibited from performing intentional inverted spins.

Gauge	Spin Indications	Remarks
Altimeter	Rapidly decreasing	May indicate up to 1000 feet above actual altitude.
AOA	2–3 units	Stalled
Airspeed	Zero	Stable
Turn Needle	Pegged in direction of spin	Spin rate: 140° per sec.
VSI	6000 fpm (pegged)	8,700 fpm
Attitude Gyro	May be tumbling	25° Nose down

Table 1-10-2 T-34C Indications of an Inverted Spin
(characterized by nose-down, upside down attitude, and negative Gs)

A flat spin is characterized by a flat attitude and transverse or "eyeball out" Gs. Since the relative wind is from directly below the airplane, the control surfaces are ineffective. The cockpit indications will be similar to an erect spin, except airspeed may vary depending on how flat the spin is. The T-34C will not enter a flat spin.

SPIN CHARACTERISTICS

During a spin, the control surfaces continue to generate forces and affect the way in which an aircraft spins.

AILERONS

Ailerons are not used to recover from a spin in the T-34C since they rarely assist in the recovery. This makes sense in that the wing is stalled and there is not much useful air going over the ailerons. Therefore, during spin recoveries, apply neutral ailerons.

On some high performance aircraft, recovery using the ailerons is necessary. The aileron on the up-going wing puts it at a higher angle of attack. This results in increased drag on the wing and produces adverse yaw to oppose the large yawing moments of the spin.

RUDDER

The rudder is generally the principal control for stopping autorotation. Due to the direction of the relative wind in a spin, the dorsal fin area or vertical stabilizer acts as a flat plate, instead of as an airfoil, with the aerodynamic force parallel to the relative wind (Figure 1-10-3). If the rudder is deflected in the same direction as the spin (pro-spin rudder), the vertical stabilizer exposed to the relative wind will be minimized. If the rudder is deflected in the opposite direction as the spin (anti-spin rudder), the vertical stabilizer exposed to the relative wind will be maximized. The drag created by the vertical stabilizer can be divided into a horizontal and vertical component. The horizontal component creates a force that opposes the yawing of the airplane. The vertical component creates a force that pulls the tail up and pitches the nose-down. Opposite rudder maximizes both of these components

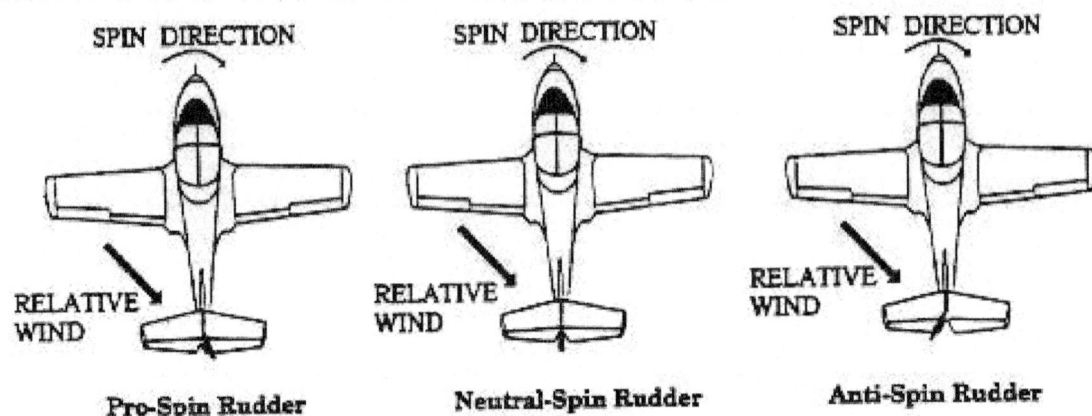

Figure 1-10-3 Rudder Forces During a Spin

The design of the vertical stabilizer and rudder and the placement of the horizontal control surfaces will significantly affect spin recovery. In the case of a swept vertical fin (Figure 1-10-4), airflow to the rudder is almost entirely blocked by the horizontal surfaces. The rudder is therefore not very effective at stopping the autorotation. With the T-34C tail design, airflow to the rudder is not blocked

The T-34C uses a dorsal fin, strakes, and ventral fins to decrease the severity of spin characteristics (Figure 1-10-5). The dorsal fin is attached to the front of the vertical stabilizer to increase its surface area. This decreases the spin rate and aids in stopping the autorotation.

Two ventral fins on the T-34C are located beneath the empennage. Ventral fins decrease the spin rate and aid in maintaining a nose down attitude. The T-34C has strakes located in front of the horizontal stabilizer. These stakes increase the surface area of the horizontal stabilizer in order to keep the nose pitched down and prevent a flat spin. These strakes changed the airflow over the nose creating an anti-rotational force.

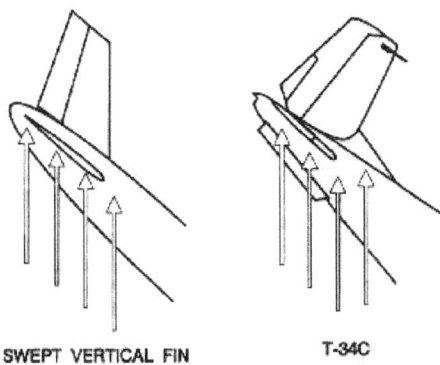

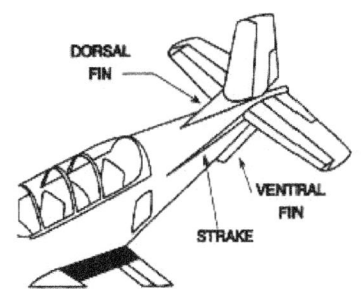

Figure 1-10-4 Effect of a Swept Vertical Fin   Figure 1-10-5 T-34C Empannage

ELEVATOR

In a stabilized spin, the horizontal stabilizer and elevator are fully stalled due to an angle of attack in excess of 50°. This results in very little lift and a great amount of drag. The drag will be maximized with full down elevator and minimized with full up elevator. This drag will also have a vertical and horizontal component like the drag of the vertical stabilizer.

Figure 1-10-6 Effect of Elevator

Rotation rate increases as the pitch attitude in a spin becomes steeper, either from the increased nose down force from the rudder or the elevator. This is due to the conservation of angular momentum. As the pitch becomes steeper, the aircraft's center of mass, which is aft of the spin axis, moves closer to the spin axis (Figure 1-10-7). The shortened moment arm necessitates an increased angular velocity in order to conserve angular momentum. This can best be shown with the ice skater analogy. As a spinning skater's arms are brought in, the spin rate increases for the same reason.

Full aft stick results in the flattest pitch attitude and therefore the slowest spin rate. Any stick position other than full aft will result in a steeper pitch attitude and an increase in rotation rate. This is referred to as an accelerated spin.

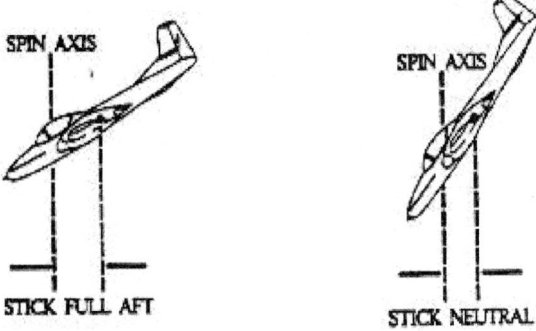

As the pitch attitude is slowly lowered, the increase in spin rate causes the center of gravity to experience a greater force away from the spin axis. The acceleration resists the airplane moving to a steeper pitch attitude. This is why brisk control inputs are emphasized in a recovery. Abrupt forward stick will drive the pitch attitude down before rotation rate can increase appreciably and build up the nose down resistance.

Figure 1-10-7 "Ice Skater" Effect

If proper recovery procedures are not followed in the T-34C, a progressive or aggravated spin could result. In a progressive spin, instead of a smooth spin recovery, the airplane will reverse spin direction. A progressive spin will result if, upon recovery, the pilot puts in full opposite rudder but inadvertently maintains full aft stick. After one or two more turns in the initial spin direction, the nose will pitch steeply down and the airplane will snap into a reversed direction of rotation. The spin reversal is disorienting and the entry more violent than with a normal erect spin. No matter how disorienting and violent the entry may be, remember to look at the turn needle to determine spin direction. To recover, apply full rudder opposite the turn needle and stick slightly forward of neutral. When rotation stops, the pilot may be mildly disoriented, so the horizon should be referenced to maintain a wings level attitude during the pull-out.

An aggravated spin in the T-34C results from pushing the stick forward while maintaining rudder in the direction of spin. Neutralizing rudder while advancing the stick may also be sufficient to enter an aggravated spin. It is essentially an extreme case of an accelerated spin. Aggravated spins are characterized by a steep nose-down pitch and an increase in spin rate. In addition, aggravated spins tend to induce severe pilot disorientation. Recovery procedures from an aggravated spin are the same as from a progressive spin.

It is interesting that although an inverted spin in a T-34C is difficult to enter and very disorienting, it is easy to recover from. This is because the entire vertical stabilizer is in an inverted position and therefore 100% exposed to the relative wind. If the pilot is hanging in the straps, he or she should be in an inverted spin. Inverted spin direction is hard to determine visually, so the turn needle should be referenced. To recover, simply apply full rudder opposite the turn needle and neutralize the aileron and elevator. The spin will recover to a steep, inverted, nose down dive. Roll or split-S out of the dive to level flight in a timely manner as airspeed will build rapidly.

FACTORS THAT AFFECT SPINS

AIRCRAFT WEIGHT

The aircraft's weight varies primarily due to fuel usage, but can also vary if items are dropped off the airplane (e.g., ordnance, fuel tanks, etc.). If an aircraft carries fuel in the wings, a large portion of the weight of the airplane is away from the center of gravity. This creates a large moment of inertia for a spin to overcome. A heavier airplane will have a slower spin entry with less oscillations due to this large moment of inertia. A lighter airplane will enter a spin more quickly, with greater oscillations possible, but will also recover from a spin faster.

PITCH ATTITUDE

The pitch attitude will have a direct impact on the speed the aircraft stalls. The higher the pitch attitude, the greater the vertical component of thrust, and the lower the stall speed. Slower stall speeds make the spin entry slower and with less oscillations. At lower pitch attitudes, the aircraft stalls at a higher airspeed and entries are faster and more oscillatory.

SPIN DIRECTION

The propeller of the T-34C is a clockwise rotating gyroscope (as viewed from behind to determine direction of rotation). If an airplane is in a right spin (nose yawing right), the nose of the T-34C will tend to pitch down due to gyroscopic precession. The T-34C will therefore have a flatter attitude when spinning to the left than to the right. This makes for smoother entries into spins that stabilize quicker. A T-34C in a right spin will have a more oscillatory entry.

T-34C SPIN RECOVERY

The spin recovery is the most positive recovery available and is 100% effective when properly applied.

Step 1. Landing gear and flaps – Check Up.

Step 2. Verify spin indications by checking AOA, airspeed, and turn needle. If recovery from erect spin does not occur within two turns after applying recovery controls, verify cockpit indications of AOA, airspeed and turn needle for steady state-spin and visually confirm proper spin recovery controls are applied.

Step 3. Apply full rudder opposite the turn needle. If positive force is not applied to maintain full deflection, spin recovery may take several additional turns.

Step 4. Position stick forward of neutral (ailerons neutral).

(1)   Erect Spin: Expect a push force of approximately 40 pounds to keep the stick forward of the neutral position.

(2)   Inverted Spin: Expect a pull force of approximately 30 pounds to place the stick in the neutral position.

Step 5. Neutralize controls as rotation stops. Also reduce power to idle to minimize altitude loss and rapid airspeed buildup.

Step 6. Recover from the ensuing unusual attitude. Aircraft will consistently recover in a steep, nose-down attitude. Slowly apply back-stick pressure until nose reaches the horizon and wings are level.

1.   Define spin. What are the two requirements for entering a spin?

2.   Compare the up-going and down-going wings in a spin with respect to AOA, C_L and C_D.

3.   What instruments should a pilot observe to determine whether he or she is in a spin?

4.   How do high gross weights affect the rate of spin entry?

5.   How do high pitch attitudes affect the rate of spin entry?

6.   What are the basic spin recovery procedures from an erect spin?
        A.   Reverse rotation, break the stall.
        B.   Stop rotation and reverse roll.
        C.   Break the stall and stop rotation.
        D.   Stop the rotation and roll.

7.   What would be the result if the rudder were maintained in the direction of spin during a spin recovery?

8.   What is the cause of an erect spin?

9.   In which direction are spin entries more oscillatory and take longer to stabilize in the T-34C?

10.   During a recovery from an inverted spin, what action should be taken when the rotation stops?

11.   What do the turn needles indicate during a spin?
        A.   Direction of spin for erect spins only
        B.   Direction of spin for inverted spins only
        C.   Centrifugal force displaces both needles outward away from the spin axis
        D.   Direction of spin always

Turning Flight

INTRODUCTION

The purpose of this lesson is to aid the student in understanding maneuver limitations as they relate to aerodynamics.

TERMINAL OBJECTIVE

Upon completion of this unit of instruction, the student aviator will demonstrate knowledge of basic aerodynamic factors that affect airplane performance.

ENABLING OBJECTIVES

1.151	Describe how the forces acting on an airplane produce a level coordinated turn.
1.152	Define load factor.
1.153	Describe the relationship between load factor and angle of bank for level flight.
1.154	State the effect of maneuvering on stall speed.
1.155	Define load, strength, static strength, static failure, fatigue strength, fatigue failure, service life, creep, limit load factor, elastic limit, overstress/over-G, and ultimate load factor.
1.156	State the relationship between the elastic limit and the limit load factor.
1.157	Describe and identify the parts of the V-n/V-g diagram, including the major axes, limit load factor, ultimate load factor, maneuvering speed/cornering velocity, redline airspeed, accelerated stall lines, and the safe flight envelope.
1.158	List and describe the phenomena that are used to determine redline airspeed.
1.159	State the limit load factors, maneuvering speed, and redline airspeed for the T-34C.
1.160	Describe the effects of weight, altitude, and configuration on the safe flight envelope.
1.161	Define asymmetric loading and state the limitations.
1.162	Define gust loading.
1.163	State what should be done not to exceed the limit load factor in moderate turbulence.
1.164	Define turn radius and turn rate.
1.165	State the effect of velocity, angle of bank, weight, slipping, and skidding on turn rate and turn radius.
1.166	Define standard rate turn.
1.167	State the approximate angle of bank for a standard rate turn in the T-34C.

Turning Flight

INTRODUCTION

Unlike an automobile or other ground supported vehicles, an aircraft can rotate about three axes. It can pitch up and down, yaw left or right, and roll to the left or the right. Because of this freedom of movement, the airplane can perform many maneuvers. However, all these maneuvers consist of turns, either horizontal or vertical, or a combination of the two. This lesson discusses turns and the limits imposed on them.

REFERENCES

1.   Aerodynamics for Naval Aviators

2.   Aerodynamics for Pilots

3.   Introduction to the Aerodynamics of Flight

4.   T-34C NATOPS Flight Manual

INFORMATION

MANEUVERING FORCES

Turning flight is described as changing the direction of the airplane's flight path by reorienting the lift vector in the desired direction. During a turn, the lift vector is divided into two components, a horizontal component (L_H) and a vertical component (L_V) (Figure 1-11-1). The horizontal component of lift, called centripetal force, accelerates the airplane toward the inside of the turn. In straight and level flight (constant altitude, constant direction) total lift is equal to weight, but in a turn, only the vertical component of the lift vector opposes weight. If the pilot does not increase the total lift vector, the airplane will lose altitude because weight will be greater than L_V. The increased lift is normally obtained by increasing the angle of attack, i.e. pulling back on the stick. As the stick moves aft, G-forces build up.

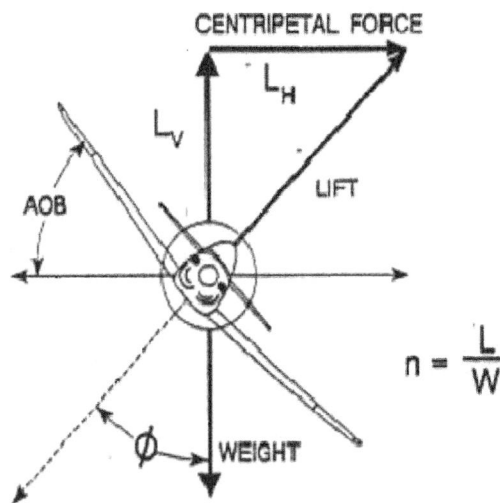

Figure 1-11-1 Turning Flight

Increasing the lift produced by the wings increases the load on the airplane. **Load factor (n)** is the ratio of total lift to the airplane's weight. It is sometimes called Gs since it is the number of times the earth's gravitational pull felt by the pilot. For example, a 3,000 pound airplane in a 60° angle of bank turn must produce 3,000 pounds of vertical lift to maintain altitude. Therefore, the wings must produce 6,000 pounds of total lift so the airplane experiences a load on its wings that is twice the force due to gravity, or 2 Gs. One "G" is what we experience just sitting or walking.

$$n = \frac{L}{W} \quad \text{or} \quad L = W \cdot n$$

In maneuvering flight, the amount of lift produced by an airplane is equal to its weight (W) multiplied by its load factor (n). By substituting $W \cdot n$ into the lift equation and solving for V, we can derive an equation for stall speed during maneuvering flight. This is called **accelerated stall speed** because it represents the stall speed at velocities greater than minimum straight and level stall speed, and load factors greater than one. Phi (φ) is the angle of bank associated with the load factor (n).

$$V_{S\phi} = \sqrt{\frac{2Wn}{\rho S C_{L\max}}} \quad IAS_{S\phi} = \sqrt{\frac{2Wn}{\rho_0 S C_{L\max}}}$$

Maneuvering the airplane will significantly affect stall speed. Stall speed increases when we induce a load factor greater than one on the airplane. Figure 1-11-2 is a generic chart that can be used for any fixed wing aircraft and assumes a constant altitude turn. It lists the load factors and percent increase in stall speed for varying angles of bank. Notice that above 45° angle of bank the increase in load factor and stall speed is rapid. This emphasizes the need to avoid steep turns at low airspeeds. An airplane in a 60° angle of bank experiences 2 Gs, but has an accelerated stall speed that is 40% greater than wings level stall speed.

A quick method for calculating accelerated stall speed is to round your normal stall speed off to a higher, round number and multiply it by the square root of the number of Gs sustained. For example, if stall speed is 92 kts and a 2 G maneuver is performed, accelerated stall speed can be estimated by rounding 92 kts up to 100 kts, then multiplying by the square root of two (1.4).

As the load factor approaches three Gs, the pilot will notice a sensation of blood draining from the head and a tendency for his or her face to sag. Further increases in G loading may cause the pilot to gray out or even temporarily lose consciousness if he or she does not correctly accomplish the anti-G straining maneuver. As the load factor approaches

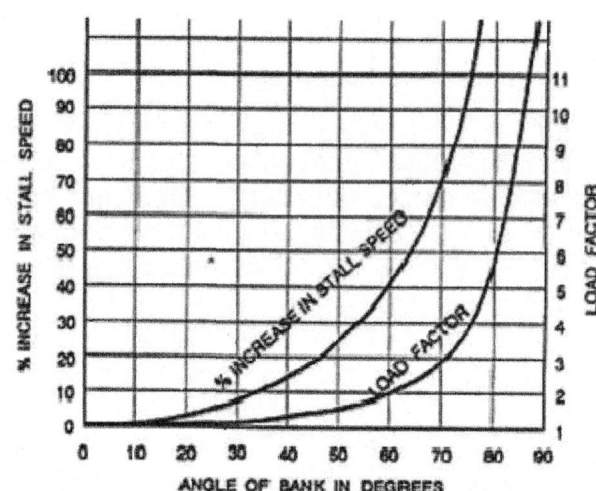

Figure 1-11-2 Stall Speed vs. AOB

negative three Gs, the pilot will notice a sensation of blood rushing to the head, and the face and eyeballs will feel like they are bulging out. Exceeding negative three Gs may cause one to "red out" or suffer from bursting blood vessels.

DEFINITIONS

A **load** is a stress-producing force that is imposed upon an airplane or component. **Strength** is a measure of a material's resistance to load. There are two types of strength: Static strength and fatigue strength. **Static strength** is a measure of a material's resistance to a single application of a steadily increasing load or force. **Static failure** is the breaking or serious permanent deformation of a material due to a single application of a steadily increasing load or force. For instance, a pencil breaks when too much force is applied and its static strength is exceeded.

Fatigue strength is a measure of a material's ability to withstand a cyclic application of load or force, i.e., numerous small applications of a small force over a long period of time. **Fatigue failure** is the breaking (or serious permanent deformation) of a material due to a cyclic application of load or force. Breaking a wire coat hanger by bending it back and forth demonstrates fatigue failure. Airplanes may experience fatigue failure on many components (landing gear struts, tailhooks and mounting brackets) due to the numerous arrested landings, catapult shots, and high G maneuvers performed in normal operation. The components are designed to withstand repeated loads, but not forever. **Service life** is the number of applications of load or force that a component can withstand before it has the probability of failing. Fatigue strength plays a major role in determining service life. Service life may apply to an individual component, or to the entire airframe.

When a metal is subjected to high stress and temperature it tends to stretch or elongate. This is called plastic deformation or **creep**. Engine turbine blades are periodically monitored for creep damage due to high heat and stress. Modern supersonic aircraft may suffer from creep damage on the skin of the airplane, especially on the leading edge of the wings.

The structural limits of an airplane are primarily due to the metal skeleton or airframe. Any time a wing produces lift, it bends upward. The wing may permanently deform if lift becomes too great. Airframe components, particularly the wings, determine the maximum load that the airplane can withstand. The two greatest loads on an airplane are lift and weight. Since weight doesn't vary greatly from one moment to the next, lift will be the force that causes the maximum load to be exceeded.

It is difficult to measure the amount of lift produced by the airplane, but it is relatively easy to measure acceleration. Since acceleration is proportional to force (Newton's Second Law), if we know the weight of the airplane, we can determine the amount of lift by monitoring the airplane's acceleration. Since load factor is a ratio of an airplane's lift to its weight, and the mass being accelerated by lift and weight is the same mass, load factor is actually the acceleration due to lift expressed as a multiple of the earth's acceleration, and can easily be measured by an accelerometer.

Structural considerations determined by the airplane's mission and desired service life force a manufacturer to meet certain limits, such as maximum load factor, airspeed and maneuvering limitations. These design limits include the limit load factor, ultimate load factor, redline airspeed and maneuvering parameters.

Limit load factor is the greatest load factor an airplane can sustain without any risk of permanent deformation. It is the maximum load factor anticipated in normal daily operations. If the limit load factor is exceeded, some structural damage or permanent deformation may occur. Aircraft will have both positive and negative limit load factors. The T-34C's limit load factor is set at +4.5 Gs and -2.3 Gs.

Overstress/Over-G is the condition of possible permanent deformation or damage that results from exceeding the limit load factor. This type of damage will reduce the service life of the airplane because it weakens the airplane's basic structure. Overstress/over-g may occur without visibly damaging the airframe. Inside the airplane are a variety of components, such as hydraulic actuators and engine mounts, which are not designed to withstand the same loads that the airframe can. Before the airframe experiences static failure these components may break if overstressed. The wing will not depart the airplane if the limit load factor is exceeded, but if an engine mount breaks, a fire could result from fuel spewing on hot engine casing. Any time an airplane experiences an overstress, maintenance personnel must inspect to determine whether damage or permanent deformation actually occurred. Always report an overstress/over-G to maintenance. Whether or not deformation or damage occurs depends on the elastic limit of the individual components.

If a rigid metal object, such as a wing, is subjected to a steadily increasing load, it will bend or twist. When the load is removed, the component may return to its original shape. The **elastic limit** is the maximum load that may be applied to a component without permanent deformation. When a component is stressed beyond the elastic limit, it will experience some permanent deformation, but may still be usable. If the force continues to increase, the component will break. To ensure the airplane may operate at its limit load factor without permanent deformation, the limit load factor is designed to be less than the elastic limit of individual components. This virtually guarantees the airplane will reach its expected service life.

Ultimate load factor is the maximum load factor that the airplane can withstand without structural failure. There will be some permanent deformation at the ultimate load factor, but no actual failure of the major load-carrying components should occur. If you exceed the ultimate load factor, structural failure is imminent (something major on the airplane will break). The ultimate load factor should be avoided since the typical airplane is rather difficult to fly after its wings tear off. The ultimate load factor is 150% of the limit load factor.

V-N / V-G DIAGRAM

The **V-n diagram** or **V-G diagram** is a graph that summarizes an airplane's structural and aerodynamic limitation. The horizontal axis is indicated airspeed, since this is what we see in the cockpit. The vertical axis of the graph is load factor, or Gs. The V-n diagram represents the maneuvering envelope of the airplane for a particular weight, altitude, and configuration.

Accelerated stall lines, or lines of maximum lift, represent the maximum load factor that an airplane can produce based on airspeed. The accelerated stall lines are determined by C_{Lmax} AOA. They are the curving lines on the left side of the V-n diagram (Figure 1-11-3). If one tries to maintain a constant airspeed and increase lift beyond the accelerated stall lines, the airplane will stall because we have exceeded the stalling angle of attack. As airspeed increases, more lift can be produced without exceeding the stalling angle of attack.

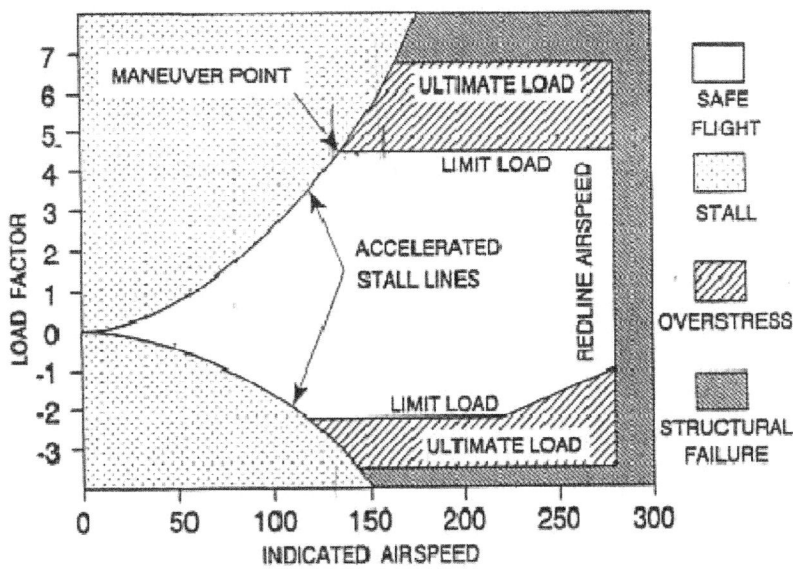

Figure 1-11-3 T-34C V-n Diagram

The limit load factors and ultimate load factors, both positive and negative, are plotted on the diagram. These lines represent the manufacturer's and the military's structural limitations. Any G load above the limit load factor will overstress the airplane. Any G load above the ultimate load factor is very likely to cause structural failure. Notice that the positive and negative limit load factors are different. Since the pilot cannot sustain a negative acceleration much greater than three Gs, the designer can save some structural weight by reducing the airplane's ability to sustain negative Gs. For this reason, most maneuvers are performed with positive accelerations.

The point where the accelerated stall line and the limit load factor line intersect is called the **maneuver point**. The IAS at the maneuver point is called the **maneuver speed (V_a)** or **cornering velocity**. It is the lowest airspeed at which the limit load factor can be reached. Below the maneuver speed, we can never exceed the limit load factor because the airplane will stall before the limit load factor is reached. The T-34C's maneuver speed is 135 KIAS at maximum gross weight.

The vertical line on the right side is called the **redline airspeed**, or V_{NE} (Velocity never-to-exceed). Redline airspeed is the highest airspeed that an airplane is allowed to fly. Flight at speeds above V_{NE} can cause structural damage. V_{NE} is determined by one of several methods: M_{CRIT}, airframe temperature, excessive structural loads, or controllability limits.

If an airplane reaches its critical Mach number (M_{CRIT}), and is not designed to withstand supersonic airflow, the shock waves generated may damage the structure of the airplane. Redline airspeed for these aircraft will be slightly below the airspeed at which they will achieve M_{CRIT}.

Redline airspeed may also be used to set limits on airframe temperature. As airspeed increases, the airplane encounters more air particles producing friction which heats up the airframe. This heating can be extreme and hazardous at high speeds. Once the temperature becomes excessive, the airframe may suffer creep damage.

Excessive structural loads may be encountered on components other than the main structural members. Control surfaces, flaps, stabilizers, and other external components are often not able to withstand the same forces that the wings or fuselage can withstand. Deflecting control surfaces at very high airspeeds may create sufficient forces to twist or break the wing or stabilizer on which they are located. Excessive horizontal stabilizer loads can be encountered in the T-34C at speeds in excess of its redline airspeed of 280 KIAS.

Controllability may determine the redline airspeed on aircraft with conventional control systems. At high airspeeds, dynamic pressure may create forces on the control surfaces which exceed the pilot's ability to overcome. Or, due to the aeroelasticity of the controls surfaces, full deflection of the cockpit controls may cause only small deflection of the control surfaces. In either case, the pilot will be unable to provide sufficient control input to safely maneuver the airplane.

FACTORS AFFECTING THE SAFE FLIGHT ENVELOPE

The portion of the V-n diagram that is bounded by the accelerated stall lines, the limit load factors and redline airspeed is called the **safe flight envelope**. The five major factors affecting the safe flight envelope are gross weight, altitude, configuration, asymmetric loading, and gust loading.

The gross weight of an airplane will affect the airplane's limit load factor and ultimate load factor. Consider an airplane whose wing is built to withstand 20,000 pounds of static load; this will determine how many Gs can be pulled. If the airplane takes off with a weight of 5,000 pounds, it could withstand 4 Gs (20,000 / 5,000 = 4). If the airplane weight decreases by burning fuel or expending ordnance, the limit load factor will increase. If the same airplane decreased its weight to 4,000 pounds, it could now withstand 5 Gs. An increase in weight will also cause the accelerated stall lines to sweep to the right since an increase in

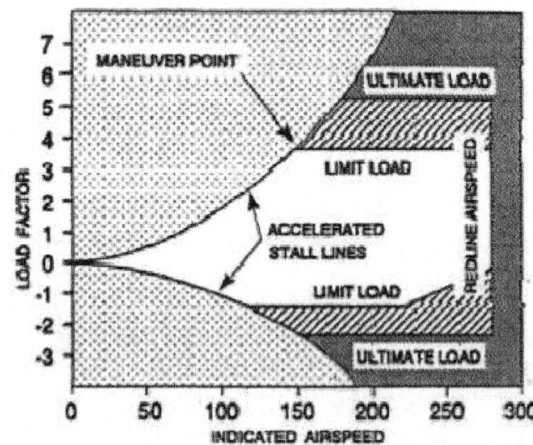

Figure 1-11-4 Effect of Increased Gross Weight

weight increases an airplane's stall speed. This causes the maneuver speed to increase (Figure 1-11-4). Weight generally does not affect redline airspeed. Since its weight changes are small compared to other aircraft, they are not accounted for in the T-34C's safe flight envelope.

As altitude increases, the speed of sound will decrease and TAS will increase for a given IAS. With an increase in altitude the indicated redline airspeed must decrease in order to keep a subsonic airplane below M_{CRIT} TAS. Since the limit and ultimate load factors are structural limits, they do not change with altitude. Since the horizontal axis is indicated airspeed, the accelerated stall lines will not change. Above 20,000 feet, the T-34C's redline airspeed decreases to 245 KIAS (Figure 1-11-5).

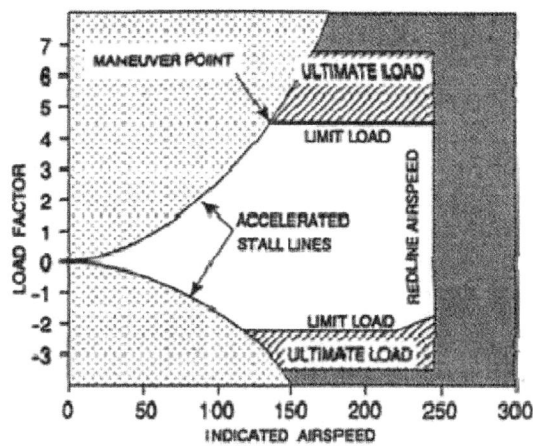

Figure 1-11-5 Effect of Increased Altitude

Another factor that affects the safe flight envelope is configuration. When the landing gear and high lift devices are extended, the envelope is substantially reduced in size. This is mainly due to the relatively weak structure of the landing gear doors and the deployed high lift devices. High airspeeds could possibly tear the landing gear doors off or bend the flaps. An airplane in the landing configuration does not need to maneuver at high speeds and create high G loading. Changing the configuration by adding external stores, such as weapons or drop tanks, may also reduce redline airspeed because the higher air loads imposed may tear them from the airplane (Figure 1-11-6).

Asymmetric loading refers to uneven production of lift on the wings of an airplane. It may be caused by a rolling pullout, trapped fuel, or hung ordnance. The V-n diagram may reflect limits that are imposed because of this condition (Figure 1-11-7). When an airplane is rolling, the up-going wing is producing more lift than the down-going wing. If the airplane performs a rolling pullout, the up-going wing may become overstressed even though the accelerometer in the cockpit shows a G load at or below the limit load factor. This would be aggravated even further if there were an imbalance of ordnance or fuel on the wings. For this reason, in all maneuvers requiring a pullout at higher than normal loading, one of the

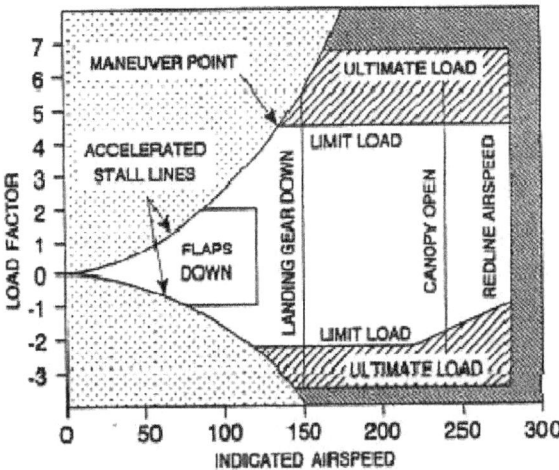

Figure 1-11-6 Effect of Configuration

first steps is to always level the wings. If the pilot were to experience an asymmetric load after a bombing run, e.g., hung ordnance, special attention must be paid to the amount of Gs and angle of bank. Because asymmetric loading is cumulative with pilot induced loading, the limit load factor due to pilot induced loads should be reduced to approximately two-thirds of the normal limit load factor. This will ensure that one wing is not overstressed. In the T-34C, the maximum load factor during asymmetric loading would be 3 Gs (2/3 X 4.5).

Gust loading refers to the increase in the G load due to vertical wind gusts. The load imposed by a gust is dependent upon the velocity of the gust. The higher the velocity, the greater the increase in load. If an airplane were generating the limit load factor during a maximum performance turn and hit a vertical gust, the gust would instantaneously increase the angle of attack of the airfoils and increase the lift on the wings enough to raise the G load above the limit load factor. For this reason, "intentional flight through severe or extreme turbulence and thunderstorms is prohibited" in the T-34C.

Vertical gusts of up to 30 feet per second may be encountered in moderate turbulence. This could produce up to 2 Gs of acceleration on the

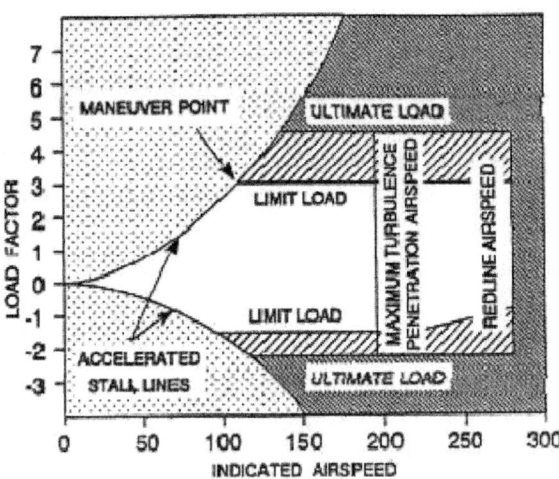

Figure 1-11-7 Effect of Asymmetric Loading or Gust Loading

airplane. Because gust loading is cumulative with pilot induced loading, the limit load factor due to pilot induced loads should be reduced to two-thirds of the normal limit load factor. Again the two-thirds rule can be used. The T-34C NATOPS recommends a limit of +3.0 Gs. Since asymmetric loads, gust loads, and pilot induced loads are all cumulative, encountering both gust loading and asymmetric loading at the same time would require the application of the two-thirds rule twice: (4.5 X 2/3) X 2/3) = 2.0 Gs. For this reason, if a pilot makes the mistake of entering a thunderstorm, she should not turn around to get out of it, but should continue to the other side because maneuvering increases the pilot induced loads.

Turbulence penetration also requires that you slow the airplane to a speed that will reduce the effects of stress caused by gust loading. NATOPS states that the maximum airspeed for the T-34C in moderate turbulence is 195 KIAS. When entering severe or extreme turbulence, the pilot should slow to an airspeed less than this. Maneuver speed is normally what the manufacturer recommends because the airplane cannot be overstressed for positive Gs at that airspeed. At a slower airspeed the aircraft will stall more easily; it makes no sense to spend more time than necessary in the turbulence.

TURN PERFORMANCE

Turn performance is measured using two different parameters, turn rate and turn radius. **Turn rate (ω)** is the rate of heading change, measured in degrees per second. **Turn radius (r)** is a measure of the radius of the circle the flight path scribes. Turn performance in a level coordinated turn is controlled only by airspeed and angle of bank. Weight, altitude, load factor, stalling angle of attack, engine performance, and wing loading may limit either the airspeed or angle of bank. This would limit maximum turn rate or minimum turn radius, however, the actual performance would still be determined using only airspeed and angle of bank. The formulas for determining the turn rate and turn radius for an airplane in coordinated flight are:

$$\omega = \frac{g \tan \phi}{V} \quad \text{and} \quad r = \frac{V^2}{g \tan \phi},$$

where ω = turn rate, r = turn radius, V = velocity, ϕ = angle of bank and g = gravity.

If velocity is increased for a given angle of bank, turn rate will decrease, and turn radius will increase. An example of this would be turning a very sharp corner on a bicycle at 5 mph versus trying to turn the same corner at 30 mph. If angle of bank is increased for a given velocity, turn rate will increase, and turn radius will decrease.

The maximum turn rate and minimum turn radius would be achieved in a 90º angle of bank turn, at the airplane's minimum velocity. However, there are limits on angle of bank and velocity. Minimum velocity, stall speed, is determined by C_{Lmax} AOA. Maximum turn performance will be achieved at the accelerated stall speed for whatever angle of bank is being flown. An increase in angle of bank increases the accelerated stall speed, and vice versa.

If an airplane's limit load is 2 Gs, the maximum angle of bank that it could maintain will be 60 degrees (Figure 1-11-2). With a limit load factor of 4.5 Gs, the T-34C is limited to about 78 degrees angle of bank in level flight.

An airplane's thrust limit may also limit its turn performance. Since induced drag is directly proportional to lift squared, an airplane pulling 5 Gs would produce 25 times as much induced drag as in level flight. If the maximum thrust available can only overcome 16 times as much induced drag, then the airplane can only maintain level flight at 4 Gs.

Of the three factors that limit turn performance, C_{Lmax} AOA and the limit load factor are found on the V-n diagram at the maneuver point. Assuming the airplane's angle of bank is not thrust limited, this is where maximum turn performance is achieved. Any deviation from the maneuver point produces an undesired result. If velocity increases at a constant load factor, turn rate will decrease and turn radius will increase. If velocity decreases at a constant load factor, the airplane will stall. If angle of bank increases at a constant velocity, the airplane will stall. If angle of bank decreases at a constant velocity, turn radius will increase and turn rate will decrease.

Turn rate and turn radius are independent of weight. Any two airplanes capable of flying at the same velocity and same angle of bank can fly in formation, regardless of their weights. The load factor and turn performance for both airplanes will be the same, although the heavier airplane will be producing more lift.

Instrument flight requires that turns be made at a standard rate. A **Standard Rate Turn (SRT)** is one in which 3º of turn are completed every second. A rough estimate used to determine standard rate turns in the T-34C is angle of bank equal to 15-20 percent of airspeed. A Standard Rate Turn in the T-34C corresponds to two needle widths' deflection on the turn needle.

COORDINATED TURNS

The turn-and-slip indicator gives the pilot a visual indication of coordinated flight. It consists of a turn needle and
a ball suspended in fluid. If the ball is centered, the
aircraft is in coordinated flight (Figure 1-11-8). If the ball
is displaced in the same direction as the turn, the aircraft
is in a slip. If the ball is displaced in the opposite direction as the turn, the aircraft is in a skid.

Whenever the aircraft becomes uncoordinated during
flight, the corrective action is to alter the amount of
rudder being used. This simply means to apply rudder in
the direction the ball is displaced. Therefore, if the ball is
displaced to the right, apply right rudder. A useful
mnemonic device for proper rudder correction is "step on

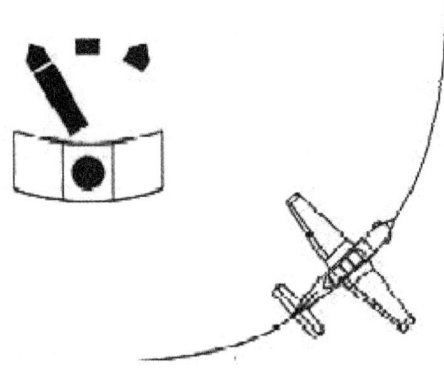

Figure 1-11-8 Coordinated Turn

the ball." The amount of rudder necessary will depend on the amount of adverse yaw.

A **skid** is caused by using too much rudder in the desired direction of turn (Figure 1-11-9). The
yawing movement is toward the inside of the turn and the balance ball is deflected toward the
outside due to centrifugal force. In a skid, turn radius will decrease and turn rate will increase.
Skids are dangerous because the airplane will roll inverted if stall occurs (a skidded turn stall).
Such a stall will probably be fatal at low altitude.

A **slip** is caused by insufficient rudder in the desired direction of turn (Figure 1-11-10). The
yawing movement is toward the outside of the turn, and the balance ball is deflected toward
the inside, due to gravitational pull. In a slip, turn radius will increase and turn rate will
decrease. Slips are useful for crosswind landings (commonly described as "wing down, top
rudder"), or when trying to increase the airplane rate of descent without increasing airspeed. A
stall while in a slip will cause the airplane to roll toward wings level (a safer reaction than in a
skid). Still, any stall at low altitude could be fatal.

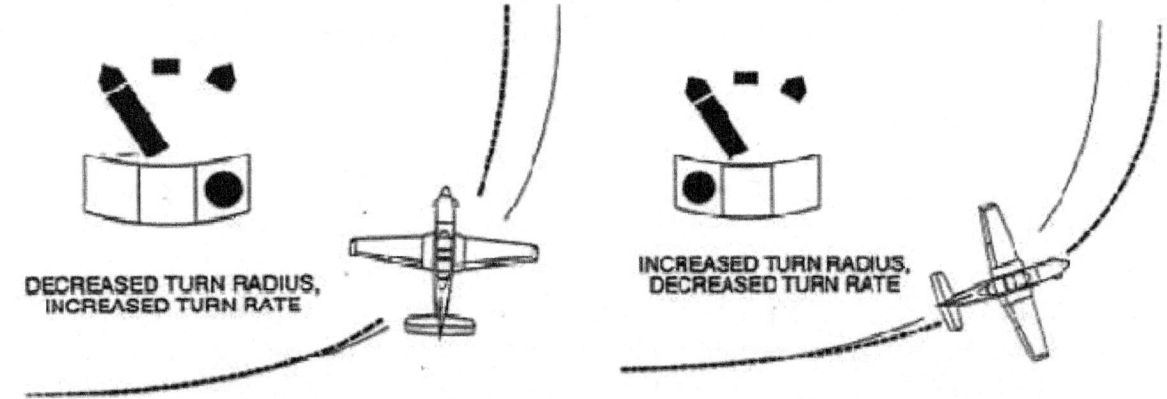

Figure 1-11-9 Skid   Figure 1-11-10 Slip

THE FINAL TURN

At this point, it is important to understand how the aircraft's turning limitations affect flying the final turn. The final turn is a critical phase of flight where disregard for aerodynamic limitations can result in disaster.

Increased Gs are required as bank angle increases during a level turn. This also applies during constant descents such as the final turn. Increased load factors result in increased stall speed (Figure 1-11-3). Many pilots have made the fatal mistake of excessively increasing bank and back pressure during the final turn in an effort to avoid overshooting the runway. They stall the aircraft, and depending upon the nature of the aircraft, the power response time, altitude, and the stall recognition point, the aircraft may not be recoverable prior to ground impact.

This situation can easily be avoided. First, pattern winds should be analyzed and the pattern planned so that an excessively tight final turn will not be required. Second, if faced with an overshooting final approach, the pilot should initiate a go-around/wave-off and continue the turn with safe amounts of bank and back pressure.

1. What must be done to the total lift vector in order to make a constant altitude turn?

2. In a constant altitude turn, what are the two components of total lift? What part does each play in turning the airplane?

3. What is the ratio of total lift to airplane weight called?

4. What determines the load factor on an aircraft in a level turn?
    A. Aircraft weight
    B. Angle of bank
    C. Aircraft speed
    D. Angle of attack

5. How many Gs will an F/A-18 produce in a 70° AOB level turn?

6. What effect does an increase in angle of bank have on the load factor in order to maintain altitude in a 90° angle of bank turn? How can an airplane maintain its altitude in a 90° angle of bank turn?

7. Describe the effects of turning/maneuvering flight on stall speed.

8. Define static strength and fatigue strength.

9. State and define the two types of metal failure.

10. Define limit load factor. What will occur when the limit load factor is exceeded?

11. What are the positive and negative limit load factors of the T-34C?

12. The _____ is the maximum load that may be applied to a component without permanent deformation. When a component is stressed beyond the _____, it will experience some permanent deformation.

13. Define ultimate load factor. What will happen if the ultimate load factor is exceeded? What is the ultimate load factor if the limit load factor is 6.0 Gs?

14. What is the vertical axis on the V-n diagram? Why is the horizontal axis of the V-n diagram labeled IAS?

15. What are the accelerated stall lines?

16. Define maneuvering speed. What is the maneuvering speed of the T-34C?

17. Define redline airspeed. List the phenomena that determine redline airspeed.

18. How does an increase in weight affect the limit load factor and ultimate load factor? Redline airspeed? Accelerated stall lines?

19. How does an increase in altitude affect the limit load factor and ultimate load factor? Redline airspeed? Accelerated stall lines?

20.   How will changing the configuration affect the size and shape of the safe flight envelope?

21.   When encountering asymmetric loading, what must the pilot do to prevent overstress? Give several examples of asymmetric loading.

22.   Identify the stress condition placed on an airplane by turbulence. What must the pilot do to prevent damage to the airplane when encountering turbulence?

23.   What is the maximum turbulent air penetration airspeed for the T-34C? If the T-34C were to accidentally fly into severe or extreme turbulence, at what airspeed should you fly?

24.   When airspeed is increased in a level turn, a constant rate turn may be maintained by _____ the angle of bank. This will _____ the load factor.

25.   As velocity decreases at a constant angle of bank, turn rate will _____. With an increase in angle of bank at a constant airspeed, turn radius will _____.

26.   What controls must the pilot adjust in order to perform a coordinated 30° AOB level turn to the left?

27.   How many seconds will it take to complete a standard rate turn for 210° of heading change?

28.   In a slip, the yawing movement is toward the (inside/outside) of the turn, turn radius will (increase/decrease), and turn rate will (increase/decrease).

29. An uncoordinated turn in which the airplane is yawing toward the inside is called a _____.
Turn radius will _____, turn rate will _____, and the turn needle and ball will be on (opposite sides/the same side).

Takeoff and Landing

INTRODUCTION

The purpose of this lesson is to aid the student in understanding takeoff and landing as it relates to aerodynamics.

TERMINAL OBJECTIVE

Upon completion of this unit of instruction, the student aviator will demonstrate knowledge of basic aerodynamic factors that affect airplane performance.

ENABLING OBJECTIVES

1.168	Define takeoff and landing speeds.
1.169	State the factors affecting the takeoff and landing speeds.
1.170	Describe the effects on true airspeed, indicated airspeed, and ground speed for takeoff and landing due to variations in weight, density, high lift devices, and wind.
1.171	Describe the forces acting on an airplane during takeoff and landing.
1.172	State the factors affecting takeoff and landing distance.
1.173	Describe the effects on takeoff and landing distance due to variations in weight, altitude, temperature, humidity, high lift devices, and wind.
1.174	Describe how crosswinds affect an airplane during takeoff and landing.
1.175	Describe how runway alignment is maintained during a crosswind takeoff or landing.
1.176	Define ground effect.
1.177	Describe the effects of ground effect on lift and drag.
1.178	State when the T-34C will be in ground effect.
1.179	State the preferred method used to stop an airplane that is hydroplaning.
1.180	State the cause of wingtip vortices.
1.181	State how interference between airplanes in flight affects the aerodynamic forces acting on each airplane.
1.182	State the airplane configuration when vortex strength is greatest.
1.183	Identify the hazards of encountering another aircraft's wake turbulence.
1.184	Identify the appropriate wake turbulence avoidance procedures.
1.185	Define wind shear.
1.186	Identify the causes of wind shear.
1.187	Identify the hazards associated with wind shear during takeoff and landing.

1.188   Describe the steps associated with wind shear recovery techniques that minimize
             the effects of wind shear on aircraft performance.

Takeoff and Landing

INTRODUCTION

Throughout aviation history many mishaps have occurred during takeoff and landing. Therefore, a firm understanding of the factors involved is very important. In addition, we will discuss the hazards of encountering wake turbulence and the appropriate wake turbulence avoidance procedures. We will then start our discussion on wind shear: its definition, its causes and the effect wind shear has on an aircraft during takeoff and landing. Finally, we will discuss techniques for recovering from a wind shear incident and methods of detecting wind shear.

REFERENCES

1. Aerodynamics for Naval Aviators

2. Aerodynamics for Pilots

3. Introduction to the Aerodynamics of Flight

4. T-34C NATOPS Flight Manual

INFORMATION

TAKEOFF AND LANDING PERFORMANCE

TAKEOFF AND LANDING SPEED

Takeoffs and landings are transitional maneuvers during which the weight of the airplane is shifted between the landing gear and the wings. The minimum airspeed for takeoff is approximately 20 percent above the power off stall speed, while landing speed is about 30 percent higher. Thus, both are affected by the same factors that affect stall speed. This safety margin minimizes operation in the region of reverse command and allows for shallow turns after takeoff, especially during an engine failure. The higher velocity on landing compensates for the decreased power setting. High lift devices are often used to decrease takeoff and landing speeds. Note that the below equations are expressed in terms of true airspeed:

$$V_{TO} \approx 1.2\sqrt{\frac{2W}{\rho SC_{L\max}}} \qquad V_{LDG} \approx 1.3\sqrt{\frac{2W}{\rho SC_{L\max}}}$$

Indicated airspeed for takeoff and landing will not be affected by changes in air density:

$$IAS_{TO} \approx 1.2\sqrt{\frac{2W}{\rho_0 SC_{L\max}}} \qquad IAS_{LDG} \approx 1.3\sqrt{\frac{2W}{\rho_0 SC_{L\max}}}$$

TAKEOFF AND LANDING FORCES

Figure 1-12-1 shows the forces acting on an airplane during takeoff or landing. During ground roll, **rolling friction (F_R)** accounts for the effects of friction between the landing gear and the runway. Like any frictional force, it is the product of a coefficient of friction (μ) and a perpendicular force (weight-on-wheels or WOW). WOW is the difference between weight and lift. The coefficient of friction is dependent upon runway surface, runway condition, tire type and degree of brake application. Note that brake application should be negligible during takeoff.

$$F_R = \mu(W - L)$$

Takeoff and landing performance are dependent upon acceleration. According to Newton's Second Law, a body accelerates in the direction of the unbalanced force acting on it. Thrust is the most out of balance force on takeoff. For an airplane to accelerate from zero to its takeoff speed, it must generate enough thrust to overcome rolling friction and drag. Although thrust and weight may change slightly during our takeoff, we will consider them to remain nearly constant.

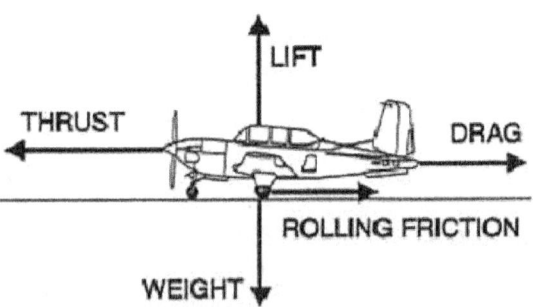

Figure 1-12-1 Takeoff and Landing Forces

As velocity increases during takeoff, the aerodynamic force increases, increasing both lift and drag (drag is primarily parasitic during a takeoff or landing). The increase in lift during the takeoff roll decreases the weight on wheels and rolling friction. $T - D - F_R$ is called the **net accelerating force**.

During the landing roll, thrust and weight still remain nearly constant (reverse thrust is discussed later). Lift and drag are functions of airspeed, so they are greatest immediately upon touchdown and decrease over the remaining landing roll. As lift decreases, weight on wheels increases causing rolling friction to increase. Drag and rolling friction will decelerate the airplane to a safe taxi speed. $D + F_R - T$ is the **net decelerating force**.

TAKEOFF PERFORMANCE

An equation for determining minimum takeoff distance is:

$$S_{TO} = \frac{W^2}{g\rho S C_{L\max}(T - D - F_R)}$$

where g = gravity.

Weight is the greatest factor in determining takeoff distance. Looking at the takeoff distance formula, we see that doubling the weight will increase the takeoff distance by a factor of four. Increasing weight requires greater lift and a higher takeoff velocity. It also increases rolling friction which decreases the net accelerating force.

Most takeoff and landing performance charts use density altitude (DA) to account for air density. Increasing DA (decreasing air density) requires a higher takeoff velocity and decreases the amount of thrust our engine can provide. This will decrease the acceleration on the takeoff roll and increase the minimum takeoff distance. There are three major factors that decrease density: increasing airfield elevation, increasing air temperature and increasing humidity. Note that takeoff indicated airspeed remains constant, regardless of temperature, humidity, and elevation.

Along with weight, these three density factors are the worst conditions for takeoff and landing. A helpful mnemonic device is the "**4-H Club**," where the members of the club are high, hot, heavy and humid. Whenever three or more of the 4-H Club are present, expect extended takeoff and landing distances. Under extreme circumstances, two or even one of the factors may cause longer takeoff and landing distances.

Using high lift devices such as flaps or BLC will decrease the takeoff distance. High lift devices decrease both the indicated and true takeoff speeds. Since true airspeed for takeoff decreases, the ground speed during takeoff will decrease, thus decreasing takeoff distance.

A headwind will decrease the takeoff distance by reducing the ground speed associated with the takeoff velocity. Conversely, a tail wind will increase takeoff distance since it increases ground speed.

LANDING PERFORMANCE

Landing is essentially the reverse of takeoff. The takeoff distance equation requires only slight modifications to be applicable to landing:

$$S_{LDG} = \frac{W^2}{g\rho S C_{L\max}(F_R + D - T)}$$

The primary consideration in landing is dissipation of the airplane's kinetic energy. Any factor affecting velocity must be considered when trying to reduce the landing distance. Final approach is flown at the lowest velocity feasible. Notice that in the landing distance equation the net accelerating forces are reversed. Drag and rolling friction are now desirable and of course, thrust is not.

An increase in weight will increase landing distance since a greater airspeed is required to support the airplane. An increase in elevation, temperature or humidity will increase landing distance since the reduced density results in a higher landing velocity. High lift devices decrease landing distance because they reduce the ground speed during the landing. A headwind reduces landing distance because it reduces ground speed. A tailwind increases landing distance since it increases ground speed. Charts for predicting takeoff and landing distance are located in the NATOPS manual for each US Navy aircraft ("Dash-1" for USAF aircraft).

The net decelerating force can be increased by use of three different techniques. **Aerodynamic braking** is accomplished by increasing the parasite drag on the airplane by holding a constant pitch attitude after touchdown and exposing more of the airplane's surface to the relative wind. This method of braking helps to reduce wear on the brakes. Drag chutes, spoilers, and speed brakes are also considered aerodynamic braking. Aerodynamic braking is used at the beginning of the landing roll. Aerodynamic breaking can also be used in flight to reduce airspeed or increase descent rates when necessary.

Mechanical braking (also called frictional or wheel braking) is effective only after enough weight is transferred to the wheels and the airplane has slowed sufficiently. A common procedure is to raise flaps or use spoilers to decrease lift and transfer the airplane's weight to the wheels when transitioning from aerodynamic to mechanical braking. Mechanical braking is used toward the end of the landing roll.

Some airplanes use reverse thrust or reverse pitch propellers (called **beta**) to shorten the landing roll. Thrust is usually negligible after touchdown, but in the case of reverse thrust or "beta" equipped airplanes, thrust increases the net decelerating force.

CROSSWINDS

Since winds do not always blow directly down the runway, the possibility of a crosswind takeoff or landing exists. The rudder is the primary means of maintaining directional control in order to compensate for the crosswind during takeoff or landing. Since the rudder loses effectiveness at low airspeeds, the self centering feature of the T-34C nosewheel provides additional directional stability if the nosewheel is contacting the runway. This enables the pilot to maintain directional control until the rudder becomes effective at higher airspeeds. The pilot must also place the ailerons into the wind during a crosswind takeoff or landing. The ailerons are not used to maintain directional control, but to overcome the lateral stability that is trying to roll the airplane away from the sideslip relative wind (crosswind).

NATOPS and the Dash-1 both contain a Takeoff/Landing Crosswind chart which allows the pilot to determine the minimum safe airspeed that the nosewheel may leave the runway during takeoff, or the minimum airspeed at which the nosewheel must return to the runway following a landing. Lifting the nosewheel below the minimum **nosewheel liftoff/touchdown (NWLO/TD) speed** may cause the airplane to weathercock or weathervane into the wind and possibly run off the runway.

Many airplanes have maximum crosswind limits that are based upon minimum nosewheel liftoff/touchdown speed. The major consideration for determining maximum authorized crosswind components is the ability to maintain directional control at low speeds. Maximum crosswind component for a takeoff or landing in the T-34C with full-flaps is 15 knots, and with no-flaps is 22 knots. For variable or gusting winds, always use the maximum wind angle and the maximum gust velocity given to determine the crosswind component.

GROUND EFFECT

A phenomenon, known as **ground effect**, significantly reduces induced drag and increases effective lift when the airplane is within one wingspan of the ground. Because takeoffs and landings are conducted at low airspeeds, induced drag makes up a large portion of the total drag on the airplane. As an airplane nears the ground, the downwash at the trailing edge of the wing is unable to flow downward. The decrease in downwash allows the total lift vector to rotate forward, increasing effective lift and decreasing induced drag. When the aircraft is one wingspan above the ground (about 33 feet for T-34C) induced drag is reduced by only 1.4%, at one-fourth the wingspan, induced drag is reduced by 23.5%, and a maximum reduction of 60% occurs just prior to touchdown or after liftoff (Figure 1-12-2).

Because of the increased lift, it is possible to get airborne at an airspeed below normal flying speed. As an airplane takes off and leaves ground effect, induced drag increases and lift decreases, which could cause an altitude loss, possibly resulting in an unintentional gear-up landing.

Entering ground effect (during landing) increases effective lift and decreases induced drag by preventing the aft inclination of the lift vector. When the plane enters ground effect it will float down the runway if the pilot does not reduce thrust. The timing of the flare and power reduction when in ground effect is the most difficult aspect of the landing phase for most students.

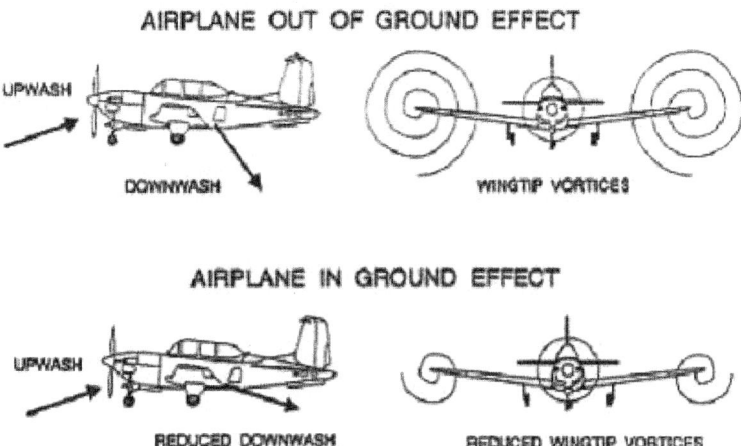

Figure 1-12-2 Ground Effect

HYDROPLANING

Hydroplaning causes the airplane's tires to skim atop a thin layer of water on a runway. If there is standing water in excess of 0.1 inches, hydroplaning may occur. Deeper tread or "channels" that allow water to escape while the tire contacts the runway may require as much as 2 inches of water before hydroplaning occurs. The speed for normal dynamic hydroplaning can be found using the following formula:

$$V_{hydroplane} = 9 \cdot \sqrt{tire\ pressure}$$

At first thought, one might think that a heavier airplane would require a faster speed before hydroplaning could occur, but experiments have shown this speed to be independent of weight. Weight only determines the size of the "footprint" that the tire makes. A heavier airplane makes a larger footprint, but the weight supported per square inch of the tire is the same. Weight has no effect on the velocity that an airplane will hydroplane, but a heavier airplane must takeoff and land at higher speeds which increases the possibility of hydroplaning.

If hydroplaning is suspected, the use of frictional brakes must be avoided, since their use may cause loss of directional control. Beta settings should be used as much as possible to slow or stop the T-34C. To minimize the effects of hydroplaning, aircrews should consider factors such as tire condition, touchdown speeds and runway condition when operating on a wet runway. Table 1-12-1 gives the tire pressure and approximate hydroplaning speeds for the landing gear of the T-34C.

	Pressure	V$_{hydroplane}$
Nosewheel	70 psi	75 kts
Main Landing Gear	90 psi	85 kts

Table 1-12-1 T-34C Hydroplaning Speeds

WAKE TURBULENCE

The spanwise airflow that moves around the wingtip does more than just create induced drag, it also creates wingtip vortices. **Wingtip vortices** are spiraling masses of air that are formed at the wingtip when an airplane produces lift (Figure 1-12-3). This disturbance is often called "jet wash" or "wake turbulence". Flying into another aircraft's vortices can lead to a variety of dangerous situations including structural damage. Vortices may instantly change the direction of the relative wind and cause one or both wings of the trailing airplane to stall, or disrupt airflow in the engine inlet inducing a compressor stall.

The most common hazard to another airplane is associated with the rolling moments that can exceed the roll control capability of the airplane. Counter control is usually effective and induced roll is least in cases where the wingspan and ailerons of the en-countering airplane extend beyond the rotational flow of the vortex. It is more difficult for airplanes with short wingspans (compared with the vortex generating airplane) to counter the imposed roll. Pilots of short wingspan airplanes, even of the high performance type, must be especially alert to vortex encounters. The most significant factor affecting your ability to counteract the roll induced by the vortices is the relative wingspan between the two airplanes.

Figure 1-12-3 Wingtip Vortices

Since vortices are a by-product of lift, they are generated from the moment an airplane rotates for takeoff until the airplane nosewheel touches down for landing. Tests show that vortices cover an area about two wingspans in width and one in height. They sink at a rate of 400 to 500 feet per minute and level off about 900 feet below the flight path of the generating airplane. Vortices will lose strength and break up after a few minutes (Figure 1-12-4). Atmospheric turbulence will accelerate this breakup. Once in contact with the ground, vortices move lateral-ly at about 5 knots. A crosswind of 4 to 6 knots may cause one vortex to remain stationary over a line on the surface, while the other vortex will move at a rate of 6 to 10 knots. This may result in the upwind vortex remaining in the touchdown zone, and the downwind vortex drifting over a parallel runway. Use caution when operating on parallel runways less than 2,500 feet apart.

The strength of a vortex depends on three main factors: airplane weight, airplane speed, and wing shape. To maintain level flight, a heavier airplane must produce more lift, and will therefore have a greater pressure differential at the wingtip where the vortex is created. A faster airplane will stretch the vortex over a longer distance. If the flaps are lowered, more lift is created at the wing root, which decreases the pressure differential at the wingtip. The greatest vortex strength occurs when the generating airplane is heavy, slow, and clean. Because weight is the most significant factor in the strength of wingtip vortices, the FAA has divided aircraft into three weight classes: Small aircraft (up to 12,500 lbs), large aircraft (12,500 to 300,000 lbs), and heavy aircraft (300,000 lbs or more). The FAA requires heavy aircraft to use the word "heavy" in all radio communications.

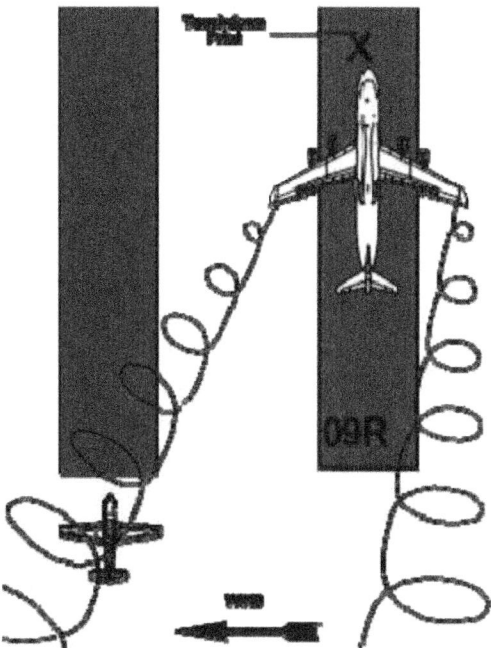

Figure 1-12-4 Lateral Vortex Movement

Flight control movements during encounters with wake turbulence tend to become instinctive, and recovery usually depends on factors other than pilot reaction. Therefore, the most important pilot technique for survival during wake turbulence is to avoid it. Prior to takeoff or landing, pilots should note the rotation or touchdown point of the preceding airplane and observe FAA separation intervals between landing and departing airplanes. Separation criteria vary from four to six miles according to the relative airplane sizes. For the Air Force, use a minimum of two minutes spacing when taking off behind a large or heavy aircraft, and use two minutes spacing minimum landing behind a large aircraft or three minutes spacing behind a heavy aircraft. To determine the actual criteria for your airplane, refer to the Aeronautical Information Manual or appropriate regulations. The following vortex avoidance procedures are recommended by the FAA.

When landing behind a larger airplane, stay at or above the larger airplane's final approach path and land beyond its touchdown point (Figure 1-12-5A). You should ensure that an interval of at least two minutes has elapsed before conducting a takeoff after a larger airplane has landed, or you can perform a "midfield takeoff" that begins beyond the larger aircraft's touchdown point (Figure 1-12-5D). If a larger airplane performs a touch-and-go or low approach, observe the same two minute interval.

When a larger airplane is departing ahead of you, ensure your landing or takeoff rotation is complete prior to the larger airplane's point of rotation (Figure 1-12-5C, D). If departing, conduct your climb-out to remain above his flight path and upwind until you turn clear (Figure 1-12-5B).

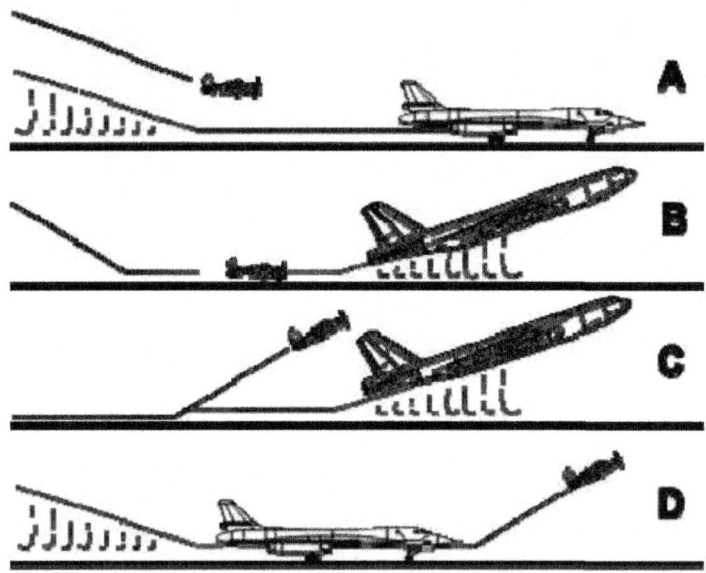

Figure 1-12-5 Vortex Avoidance

You should be aware of another hazard since it occurs well clear of the runway. Whenever a helicopter is in a hover, tremendous amounts of rotor downwash are produced. For this reason, small airplanes should avoid operating within three rotor diameters of any hovering helicopter. In forward flight, treat any helicopter as you would an airplane of similar size and weight.

During formation flying and in-flight refueling, airplanes close to one another produce a mutual interference especially when the trailing airplane is slightly aft and below the lead airplane. The leading airplane experiences an effect that is similar to ground effect because of a reduction in downwash and induced drag. For the second airplane, this mutual interference of the flow pattern can instantaneously alter the direction of the relative wind that the airfoils are sensing. Flying through lead's flightpath will place you in wake turbulence, which could result in an over-G or a flameout.

WIND SHEAR

Wind shear is defined as a sudden change in wind direction and/or speed over a short distance in the atmosphere. Wind shear is most often caused by jet streams, land or sea breezes, fronts, inversions and thunderstorms. When we discuss wind shear we are really talking about the boundary between bodies of air which have different wind characteristics. These boundaries of air can exist in the vertical or horizontal plane and vary in intensity. Weak shears distribute the wind change from one body of air to the other gradually. A pilot flying through this type of shear may not even notice a change in aircraft performance. On the other hand, strong shears distribute the wind abruptly creating rapid changes in aircraft performance.

Wind shears can be very complex combinations of wind velocities. Usually the more complex the wind shear, the more difficult it is for the pilot to react correctly. To simplify things we will limit our discussion in this section to horizontal wind shears so that we may gain a basic understanding of how wind shear will affect aircraft performance.

Wind shears change airflow over the aircraft. The velocity of the relative wind can be altered causing immediate changes in the indicated airspeed and/or angle attack of the aircraft. Once the aircraft is stabilized in the body of air it behaves as if nothing happened. You have probably experienced this effect while riding on a moving sidewalk or escalator. As you step onto these moving surfaces you feel a little unstable for a few seconds. Shortly thereafter, you stabilize and function normally. The only difference is your "groundspeed" is now a little faster.

If we confine ourselves to the horizontal plane, we can say wind shear either causes an increases or decrease in aircraft performance. With ample airspeed and altitude, wind shear does not pose a serious threat. However during slow airspeed and low altitude operations, such as during takeoffs and landings, wind shear becomes hazardous.

WIND SHEAR DURING TAKEOFF

Increasing Performance Wind Shear: Figure 1-12-6 shows an aircraft passing through a wind shear which increases indicated airspeed by 20 knots. The increase in IAS results in an increase in lift and therefore causes an initial increase in performance. During takeoff we maintain a constant attitude until reaching climb airspeed. Therefore, as long as a proper climb attitude is maintained, a wind shear with a headwind component on takeoff does not pose a serious threat.

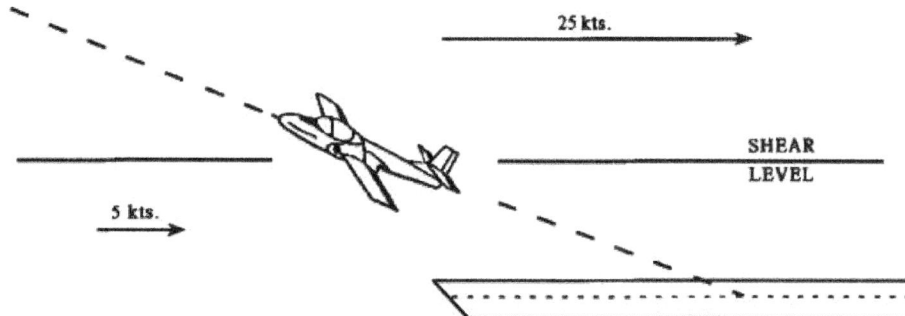

Figure 1-12-6 Increasing Performance Wind Shear on Takeoff

Decreasing Performance Wind Shear: Figure 1-12-7 shows an aircraft entering a shear which decreases the indicated airspeed by 25 knots. This causes a significant decrease in performance. If you were just about to raise the gear at 100 knots your airspeed would drop to 75 knots as the shear was entered. A rapid drop of airspeed will require great skill to maintain aircraft control. An increase in angle of attack in this situation will probably result in an approach to stall indication and possibly a stall. This example illustrates how shear can be dangerous during a decreasing performance wind.

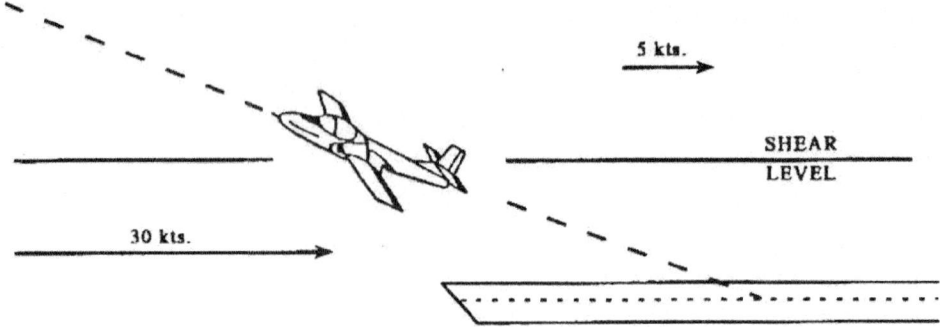

Figure 1-12-7 Decreasing Performance Wind Shear on Takeoff

WIND SHEAR DURING LANDING

An aircraft established on a glide path for landing is usually trimmed for a constant airspeed descent. Any change in indicated airspeed will cause a change in pitch due to trim and a change in the rate of descent. The pilot will have to make some control inputs to maintain the desired glide path.

Increasing Performance Wind Shear: Figure 1-12-8 shows an aircraft descending through a shear which increases its indicated airspeed. Notice that the transition from a tailwind to zero wind causes an increase in performance. This shear causes the aircraft to pitch up and rise above the glidepath. The pilot counters this by reducing the power and lowering the nose. However, the pilot may over correct and descend below the glidepath. Once back on glidepath, a higher power setting will be required to compensate for the slower ground speed (new rate of descent) within the new body of air.

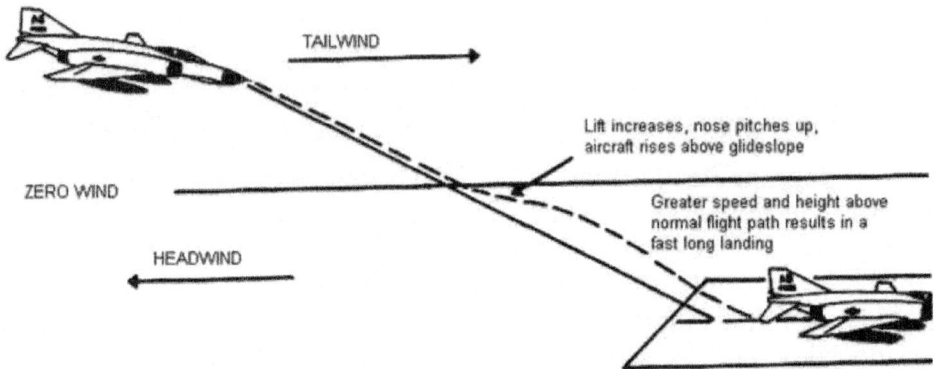

Figure 1-12-8 Increasing Performance Wind Shear on Landing

For the purpose of comparing relative power settings, assume a hypothetical thrust scale of from 1 to 10, and that the normal no-wind power setting is 6. Because of the tailwind (higher groundspeed) above the shear, the pilot needs a power setting of 5 to maintain the glidepath. When the aircraft crosses the shear and begins to rise above glidepath, the pilot reduces power to 4. As the aircraft returns to glidepath, a power setting of 6 will be required to maintain the glide path. This is due to the lower rate of descent required by a slower groundspeed. Notice that a higher power setting is required after the shear than before the shear to maintain the glidepath. In other words, you must eventually add more power than was removed to stabilize on the glidepath.

The above example assumed the shear level was high enough above the ground for the pilot to reestablish the glide path. However, the shear might be low enough to cause a short landing (Figure 1-12-9).

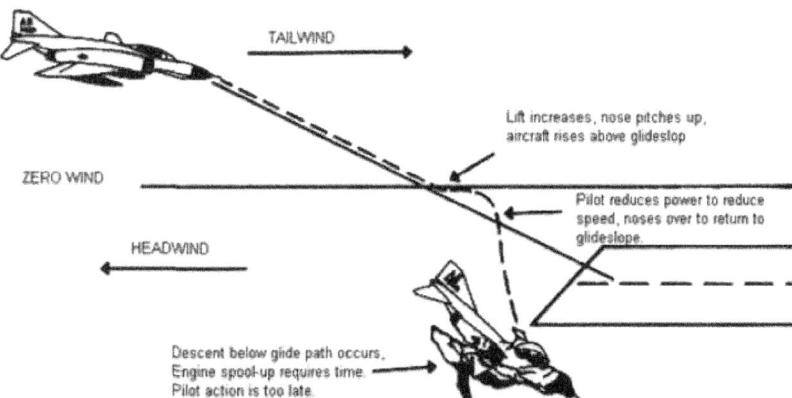

Figure 1-12-9 Increasing Performance Wind Shear on Landing Resulting in Short Landing

Decreasing Performance Wind Shear: Figure 1-12-10 shows an aircraft descending through a shear which decreases its indicated airspeed. This shear causes the aircraft to pitch down and descend below the glide path. The pilot counters this by adding power and raising the nose. However, the pilot may over correct and rise above the glidepath. Once back on glide path, a lower power setting will be required to compensate for the higher ground speed and new rate of descent within the new body of air. Again assuming a thrust scale of from 1 to 10, we can examine the power requirements during this type of shear.

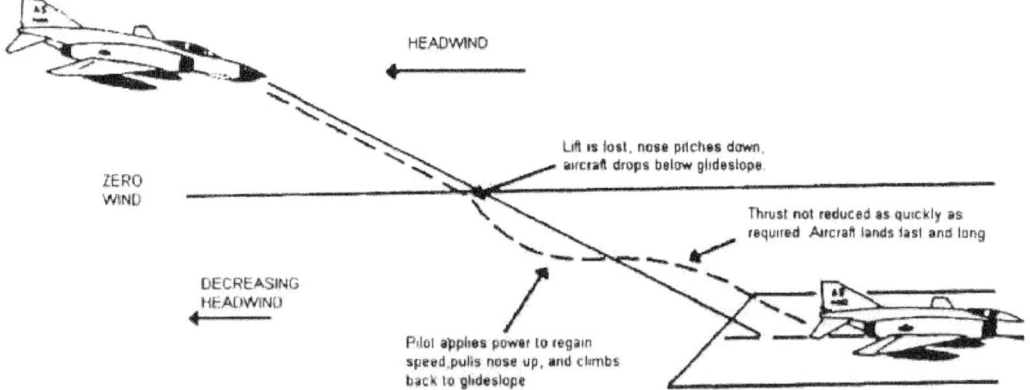

Figure 1-12-10 Decreasing Performance Wind Shear on Landing

Let's assume the normal no-wind power setting of 6. Because of the headwind above the shear, the pilot needs a power setting of 7 to maintain the glidepath. Then the aircraft crosses the shear and begins to descend below glide path, the pilot increases power to 8 or 9. As the aircraft returns to glide path, a power setting of 6 will be required to maintain it. Again, this is because of the higher rate of descent dictated by a higher groundspeed. The point to remember is that you must eventually reduce power by more than the amount added to stabilize on the glide path.

If a strong decreasing performance wind shear is encountered at very low altitude, a pilot may have insufficient time and power to overcome the resulting loss of lift (Figure 1-12-11). The outcome will invariably be a crash short of the runway.

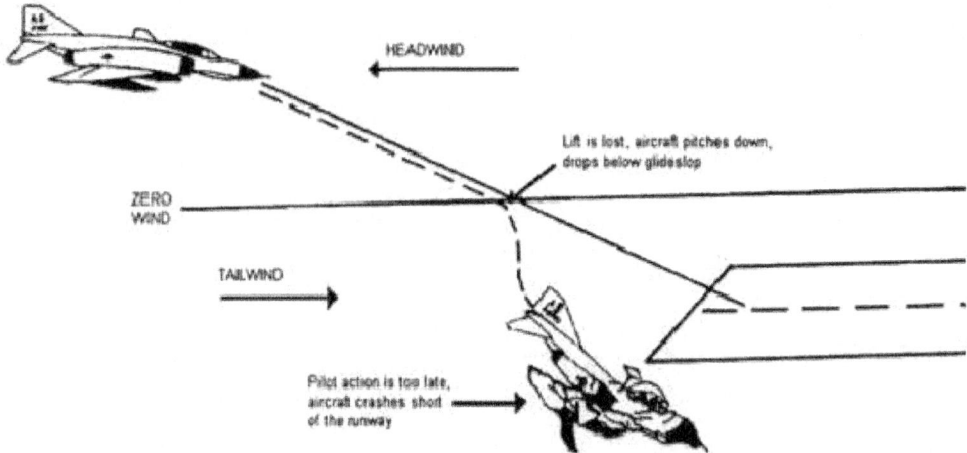

Figure 1-12-11 Decreasing Performance Wind Shear on Landing Resulting in a Crash

WIND SHEAR AVOIDANCE

The best technique for dealing with wind shear is to avoid it. If a moderate to strong wind shear is expected, delay your takeoff or landing until the shear condition no longer exists. Anytime wind shear is experienced, pilots should consider going around. If airborne and unable to delay, consider diverting to a place with more favorable conditions. Strong shears like those associated with microburst activity must be avoided. An aircraft encountering a microburst can experience significant increases followed by decreases in indicated airspeed. These can occur in a relatively short period and severely impair an aircraft's ability to maintain controlled flight.

It is important to give a PIREP (Pilot Report) after any wind shear encounter. When able, report the net change in indicated airspeed that resulted from the shear.

METHODS OF WIND SHEAR DETECTION

Because wind shear is such a dangerous phenomenon, early detection is vital to mishap prevention. In most wind shear accidents there have been warning signs that were ignored, misinterpreted or misunderstood. You must evaluate the warning signs and make a decision quickly an decisively. Here are some very important clues that indicate the presence of microburst.

Visual cues are very important because they help substantiate the information given by the weather briefer. In fact, in many fatal wind shear mishaps the pilot continued the approach or takeoff in visible and known thunderstorm conditions. Visual cues include virga, localized blowing dust (especially in circular or elliptical patterns), rain shafts with rain diverging away from the core of the cell, and of course an indication of lightning or tornado-like activity.

Wind Shear Alert Systems are another source of information about potential wind shear activity. There are several types in operation today at many civilian fields, especially those with a history of strong winds. For example the Low Level Wind Shear Alert System (LLWAS)

measures the wind speed and direction at several points on the ground and compares them with a reference sensor located near the center of the airfield. Because of the small diameter of microbursts, many may go undetected. When they are detected, they are on the field and it may be too late. There are also some Doppler radar systems which show greater accuracy in wind shear warnings. A NEXRAD Doppler radar system is a ground based radar that can very accurately detect microburst activity. Systems onboard modern aircraft monitor changes in wind velocity and aircraft acceleration to provide wind shear warning and pitch guidance to help escape wind shear. Unfortunately, these systems are not on our trainer aircraft.

PIREPS and Weather Alerts are one of the best sources of wind shear information. If you encounter wind shear, it is imperative that you make a PIREP to approach control or tower so they may notify other aircrews. Your PIREP should include the location where the shear was encountered, an estimate of its magnitude and most importantly a description of what was experienced, such as turbulence, airspeed gain or loss, glidepath problems, etc.

Departure or, arrival weather reports calling for gusty winds, heavy rain or thunderstorms should be a clue that a high potential for microburst activity exists. When you receive your pre-flight briefing or call ahead for an en route update, be alert for any convective activity that might be building. It is important to remember that weather changes very quickly. The best briefing may not prepare you for every situation that you may encounter.

It is important to understand that each piece of evidence is cumulative and as more indicators become present the potential for microburst wind shear activity increases.

1.   Indicate the effect of each of the following factors on takeoff and landing true and indicated airspeeds.

	V_{TO}	V_{LDG}	IAS_{TO}	IAS_{LDG}
Weight ↑				
Altitude ↑				
Headwind				
High Lift Devices				

2.   State the forces acting on an airplane during takeoff roll. Which one is most unbalanced?

3.   What is rolling friction? What variables determine rolling friction and the coefficient of friction? When is rolling friction greatest?

4.   What are the factors that affect takeoff and landing?

5.   What variable has the single greatest effect on the minimum distance required to takeoff or land?

6.   What is the effect of an increase in altitude on takeoff distance? Temperature?

7.   What will an increase in humidity do to landing distance?

8.   If an airplane takes off with a tailwind, takeoff distance would _____, true airspeed would _____ and ground speed would _____.

9.   What are the worst conditions for takeoff?

10.   Using a drag chute to increase drag is an example of _____ braking.

11.   Given the following runway winds, check the types of landings that are allowed in T-34C.

Runway	Headwind	Crosswind	Full Flap	No Flap
A	25	0		
B	21.5	12.5		
C	12.5	21.5		
D	0	25		

12.   What is the effect of ground effect on total lift, effective lift, induced drag, total drag, and thrust required?

13.   With _____ inches of standing water on the runway, a pilot should utilize Beta or aerodynamic braking to slow the airplane prior to wheel braking. This procedure will greatly reduce the risk of _____.

14.   What hazard to flight is the result of the increase in downwash caused by spanwise flow around the wingtip on a finite wing?

15.   Wingtip vortices will cover an area approximately _____ wide and _____ high. They will be strongest when the generating aircraft is _____, _____, and _____.

16.   When taking off behind a heavy aircraft in a T-34C, the pilot should allow for a minimum spacing
        of _____ minute(s).

17.   What should a pilot do to avoid a larger airplane's wake turbulence during landing?

18.   What should a pilot do to avoid a hovering helicopter's wake turbulence during ground taxi
        operation?

19.   What are telltale signs of microburst activity in the vicinity of thunderstorms?

20.   Tower informs a pilot waiting number one for takeoff that a landing aircraft just reported a 20
        knot decrease in airspeed due to wind shear at approximately 100 feet AGL. What should the
        pilot do?

21.   How can a pilot determine whether a wind shear condition exists?

Glossary

absolute altitude The aircraft's height above the terrain directly beneath the aircraft, measured in feet above ground level (AGL). Absolute altitude is found by subtracting the terrain elevation from the true altitude.

absolute ceiling The maximum altitude above sea level in a standard atmosphere that an airplane can maintain level flight.

accelerated spin A spin in which the control stick is not held in the full aft position. An accelerated spin is characterized by steeper pitch attitudes and higher spin rates.

accelerated stall A stall in which the load factor is greater than one, as in a pullout. Usually more violent and disorienting than a normal stall.

accelerated stall line A curved line describing the number of g's that can be generated at a given indicated airspeed as a function of $C_{L_{MAX}}$ angle of attack for a particular airfoil. Also called line of maximum lift.

acceleration A change in the velocity of a body with respect to magnitude or direction, or both.

accelerometer An instrument that measures one or more components of the acceleration of a vehicle.

adverse pressure gradient A pressure gradient of increasing static pressure in the direction of airflow.

adverse yaw Yaw in the opposite direction of aileron roll input.

aerodynamic balance The feature of a control surface that reduces the magnitude of the aerodynamic moment around the hingeline. See shielded horn.

aerodynamic braking A technique for slowing an airplane to a speed suitable for frictional braking. Aerodynamic braking is accomplished by increasing the surface area exposed to the relative wind in order to increase parasite drag, primarily by holding the nose of the airplane in the landing attitude.

aerodynamic center (AC) The point along the chordline of an airfoil where all changes in aerodynamic force effectively take place. It is normally located at the point of 25% chord.

aerodynamic force (AF) A force acting on an airfoil that is the result of air pressure and friction distribution over the surface of the airfoil.

aerodynamics The science that studies the motion of gaseous fluid flows, and of their actions against and around bodies, and of the forces acting on bodies within that flow.

aerodynamic twist Form of wing tailoring that employs a decrease in camber and/or relative thickness from wing root to wingtip. The wing root is more positively cambered and/or thicker (relative to the chord) than the tip, resulting in a root first stall pattern. Also called section variation.

aileron A movable control surface, attached to the wing of an airplane, used to produce a rolling moment around the longitudinal axis by creating unequal lifting forces on opposite sides of an airplane.

aileron reversal Reversal of the control effect usually produced by an aileron, caused by a moment around the aerodynamic center twisting the wing and changing its angle of attack.

aircraft (A/C) Any device used or intended to be used for flight in the air.

airflow A flow or stream of air. A rate of flow measured by mass per unit time.

airfoil A streamlined shape designed to produce lift as it moves through the air.

airframe The structural components of an airplane including the framework and skin of such parts as the fuselage, wings, empennage, landing gear, and engine mounts.

airplane An engine driven, heavier-than-air, fixed-wing aircraft that is supported by the dynamic reaction of airflow over its wings.

altimeter Any instrument for measuring altitude. An instrument similar to an aneroid barometer that uses the change of atmospheric pressure with altitude to indicate the approximate elevation above a given reference.

altitude The height of a point, measured from a reference plane, such as mean sea level.

ambient Pertaining to the air or air conditions around a flying aircraft but undisturbed or unaffected by it.

aneroid barometer An instrument for measuring the pressure of the atmosphere which operates on the principle of having changing atmospheric pressure bend a metallic surface which, in turn, moves a pointer across a scale graduated in units of pressure.

angle of attack (AOA, α) The angle formed between the relative wind and the chordline of the airfoil

angle of bank (AOB, ϕ) The angle between the horizon and the lateral axis of an aircraft. The angle of lateral displacement (roll) of an aircraft, especially in making a turn

angle of climb (AOC, γ) The angle between the horizon and the flightpath of a climbing aircraft.

angle of descent (γ) The angle between the horizon and the flightpath of a descending aircraft.

angle of incidence The angle between the airplane's longitudinal axis and the chordline of its wing. The root chord is commonly chosen to measure the angle of incidence.

angular acceleration Rate of change of angular velocity.

anhedral angle A negative dihedral angle. Also called cathedral angle.

approach A specified flightpath and associated altitudes to be flown in preparation for a landing, especially a published instrument approach.

artificial feel A method of simulating, altering, or otherwise enhancing the feedback or control feel that is transmitted to the cockpit controls by the forces acting on the control surfaces.

aspect ratio (AR) The ratio of the wingspan to the average chord.

attitude The orientation of an aircraft as determined by the relationship between its axes and some reference line or plane. Usually refers to nose attitude or pitch attitude.

automatic slot High lift device that consists of a movable vane attached to the leading edge of the wing that moves away from the body of the wing to allow airflow from below the wing to reach the upper surface and reenergize the boundary layer, delaying boundary layer separation. See slat.

autorotation During a spin, a combination of roll and yaw that is self sustaining.

average chord (c) The geometric average of every chord from the wing root to the wingtip. Also called mean geometric chord.

axis A reference line passing through a body, around which the body rotates.

axis system A set of three mutually perpendicular axes, intersecting at the center of gravity of an aircraft, around which the motions, moments, and forces of roll, pitch, and yaw are measured.

bank The position or attitude of an aircraft when its lateral axis is inclined from the horizontal.

Bernoulli's Equation $P_T = P_S + q$. (After Daniel Bernoulli, 1700-1782, Swiss scientist.) In aerodynamics, a law or theorem stating that in a flow of incompressible fluid, the sum of the static pressure and the dynamic pressure along a streamline is constant if gravity and frictional effects are disregarded.

boundary layer The layer of airflow over the surface of an airfoil, which shows local airflow retardation caused by viscosity. The boundary layer is very thin at the leading edge of an airfoil (about 1 mm) and grows in thickness as it moves over a body. It is composed of laminar flow and turbulent flow.

boundary layer control (BLC) The control of the airflow within the boundary layer in order to prevent its separation at high angles of attack. See also slot and slat.

buffeting The beating, shaking, or oscillation of an aircraft's structure or surfaces by an unsteady flow, gusts, turbulence, etc.

cabin Compartment of an aircraft in which passengers, troops, or cargo are loaded.

calibrated airspeed (CAS) Indicated airspeed corrected for instrument error.

calibrated altitude Indicated altitude corrected for instrument error.

camber The curvature of the mean line of an airfoil from leading edge to trailing edge; the amount of this curvature.

cantilever A beam or object supported only at or near one end, or one point; without external bracing.

cathedral See anhedral.

center of gravity (CG) The point at which the weight of an object is considered to be concentrated.

chord A measure of the chordline from the leading edge to the trailing edge of an airfoil. The chord may vary in length from the wingtip to wing root. The root chord, c_R, is the chord at the wing centerline and the tip chord, c_T, is measured at the wingtip.

chordline An infinitely long, straight line drawn through the leading and trailing edges of an airfoil.

chordwise flow Airflow perpendicular to the leading edge of an airfoil; airflow along the chord of an airfoil. Since chordwise flow is accelerated over a wing, it produces lift.

cockpit Compartment of an aircraft in which the flight crew, especially the pilot(s), are located. The cockpit is where the aircraft is controlled from.

coefficient of aerodynamic force (C$_F$) The dimensionless portion of the aerodynamic force that is a function of angle of attack, camber, aspect ratio, compressibility, and viscosity.

coefficient of drag (C$_D$) The dimensionless portion of the total drag on an airfoil that is dependent on the same variables that affect C_F.

coefficient of friction (μ) A dimensionless number whose value depends primarily on the type of material and condition of the two surfaces that are in contact.

coefficient of lift (C$_L$) The dimensionless portion of the total lift on an airfoil that is dependent on the same variables that affect C_F.

compressibility The property of a substance that allows its density to increase as pressure increases.

compressible flow Flow at speeds sufficiently high that density changes in the fluid can no longer be neglected.

constant-speed propeller A propeller designed to maintain engine speed at a constant RPM, automatically increasing or decreasing pitch as engine speed tends to increase or decrease.

continuity equation $\rho_1 A_1 V_1 = \rho_2 A_2 V_2$. Principle of physics that states that for fluids, the mass flow rate has the same value at every position along a closed tube.

control feel The feel or impression of the stability and control of an aircraft that a pilot receives through the cockpit controls, either from aerodynamic forces acting on the control surfaces or from devices simulating these aerodynamic forces.

control force A force, either aerodynamic or pilot induced, acting on a control surface.

control horn A short lever or rigid post attached to a control surface, to which a control cable, wire, line, or rod is attached.

controllability The capability of an aircraft to respond to control inputs, especially in direction or attitude.

control stick A lever for controlling the movements of an aircraft in flight. On a fixed-wing airplane, the control stick operates the elevators by a fore-and-aft movement and the ailerons by a side-to-side movement.

control surface A movable airfoil or surface, such as an aileron, elevator, rudder, or spoiler used to control the attitude or motion of an airplane and to guide it through the air.

cosine (cos) In a right triangle, the function of an acute angle that is the ratio of the length of the adjacent side to the length of the hypotenuse.

creep The gradual reduction in a material's strength over time due to high temperature and stress. Also known as plastic deformation.

critical altitude The maximum altitude at which, in the standard atmosphere, an engine produces its sea level rated horsepower or torque.

critical Mach number (M_{CRIT}) The free airstream Mach number that produces the first evidence of local sonic flow.

crosswind A wind blowing across the flightpath of an airplane.

density (ρ) Mass per unit volume.

density altitude (DA) Density altitude is pressure altitude corrected for nonstandard temperature. Density altitude is the pressure altitude on a standard day that has the same density as the ambient air.

dihedral angle The angle between the spanwise inclination of a wing and the lateral axis. It is the upward slope of the wings when viewed from head on. A negative dihedral is called anhedral.

directional control Control of the longitudinal axis around the vertical axis; yaw control.

directional divergence A departure from equilibrium around the vertical axis caused by negative directional static stability. Condition of flight in which the reaction to a small initial sideslip is an increase in sideslip angle. This would result in the airplane yawing broadside to the relative wind.

directional moment A moment created around an aircraft's vertical axis.

directional stability The stability of an aircraft around its vertical axis. The reaction of an aircraft to a sideslip.

dive A steep descent, usually power on.

downwash Chordwise airflow from the upper surface of an airfoil passing downward behind the trailing edge to the lower surface. Downwash decreases the amount of lift produced by the wing. Any downward moving airflow.

drag (D) That component of the aerodynamic force acting parallel to, and in the same direction as the relative wind. It acts as a retarding force.

Dutch roll Dynamic stability that is the result of strong lateral and weak directional static stability. An airplane prone to Dutch roll would appear to describe a figure eight on the horizon and would tail wag.

dynamic pressure (q) The pressure of a fluid resulting from its motion, equal to one half the density times the velocity squared ($q=1/2\rho V^2$).

dynamic stability The oscillatory motion of a body, beyond its initial tendency to move toward or away from equilibrium, after a disturbance. A measure of displacement with respect to time.

elastic limit The maximum load that may be applied to a component without permanent deformation.

elevator A control surface, attached to a horizontal stabilizer that produces a pitching moment around the airplane's lateral axis.

empennage The assembly of stabilizing and control surfaces at the tail of an airplane.

endurance The length of time that an aircraft can fly under specified conditions without refueling.

energy The ability or capacity to do work, expressed in foot-pounds.

equilibrium Flight condition that exists when the sum of the forces and moments acting around the center of gravity equal zero. The absence of linear or angular acceleration.

equivalent airspeed (EAS) The true airspeed at sea level on a standard day that produces the same dynamic pressure as the actual aircraft condition. It is equal to calibrated airspeed corrected for the compressibility of air.

equivalent parasite area (f) The total surface area of an airplane that contributes to parasite drag. Normally less than cross sectional area due to the effects of streamlining.

erect spin A spin characterized by positive g's and an upright attitude.

fatigue failure The breaking (or serious permanent deformation) of a material due to a cyclic application of load or force.

fatigue strength A measure of a material's resistance to a cyclic application of load or force.

feathered propeller A propeller whose blades have been rotated so that the leading and trailing edges are nearly parallel with the aircraft flightpath to minimize drag and to stop propeller rotation.

feedback The transmission of forces initiated by aerodynamic action on control surfaces to the cockpit controls. The actual forces transmitted to the cockpit controls.

fence A stall fence.

final / final approach That portion or leg of an approach pattern after the last turn, in which the aircraft is in line with the runway in the landing direction.

finite wing A wing with a finite span; a wing with wingtips.

fixed slot A slot that remains open at all times.

flap A high lift device consisting of a hinged, pivoted, or sliding airfoil or plate, or a combination of such objects regarded as a single surface, extended or deflected for increasing camber. Used primarily to decrease the takeoff or landing velocity.

flat spin A spin characterized by transverse g's and an attitude flatter than an erect spin.

flightpath (FP) The path described by an airplane's center of gravity as it moves through an air mass.

flow separation The breakaway of the boundary layer airflow from a surface; the condition of a flow separated from the surface of a body and no longer following its contours.

flutter A vibration or oscillation of a control surface or wing created and maintained by aerodynamic forces and the elastic and inertial forces of the object itself.

force A vector quantity equal to the push or pull exerted on a body. By Newton's Second Law, a force is a function of an acceleration and the mass of the body.

form drag Drag resulting from airflow over a surface with some frontal area, often referred to as pressure drag, profile drag, or plate drag.

fowler flap A high lift device that consists of a sliding airfoil attached to the trailing edge of a wing that increases camber, wing area, and uses BLC to increase the C_L.

friction Resistance due to the rubbing of one body or substance against another. Air friction results from the viscosity of the air, or its tendency to stick to a surface.

friction drag Drag arising from friction forces at the surface of an aircraft, due to the viscosity of the air.

fuel flow The rate of fuel being consumed by an aircraft's engine.

fuselage The main structural component of an airplane.

full-cantilever Supported at one point only, as in a full-cantilever wing, or a wing that is entirely internally supported, with no external bracing.

G (gravitational acceleration) A constant, equal to 32.2 ft/sec^2, representing the acceleration on an object due to the Earth's gravity.

General Gas Law $P = \rho R T$. Law of physics that shows the relationship between properties of air: pressure (P), density (ρ), and temperature (T). R is a constant for any given mixture of gases (such as dry air).

geometric twist Form of wing tailoring that employs a decrease in the angle of incidence from the wing root to the wingtip. The wing root has a higher angle of incidence than the wingtip, causing it to stall first.

glide A shallow descent, usually associated with power off flight.

glide endurance (GE) The maximum time that an airplane can stay airborne in a glide as a function of weight, altitude, and angle of attack.

glide range (GR) The maximum distance that can be traveled in a glide as a function of altitude, wind, and lift to drag ratio.

glide ratio The ratio of the horizontal distance traveled to the vertical distance descended in a glide. Glide ratio is equal to the lift to drag ratio.

gross weight The total weight of a fully loaded aircraft.

ground effect The dramatic reduction of induced drag and thrust required that occurs within one wingspan of the ground or other surface.

ground speed (GS) An airplane's actual speed over the ground.

gust A sudden and brief change of wind speed or direction.

gust load A load imposed upon an aircraft or aircraft member by a gust.

gyroscopic precession The resultant action or deflection of a spinning disc when a force is applied parallel to its axis. The resultant force occurs $90°$ ahead in the direction of rotation, and in the direction of the applied force.

headwind A wind blowing from directly ahead, or blowing from a forward direction such that its principal effect is to reduce ground speed.

helicopter A rotorcraft that, for its horizontal motion, depends principally on its engine driven rotors.

high lift device Any device, such as a flap, or boundary layer control device, used to increase the lift of a wing by increasing the C_L or area of the wing. The result is a reduction of takeoff and landing speeds. Increases in C_L are achieved by increasing the camber of an airfoil, or by controlling the kinetic energy in the boundary layer.

hingeline The transverse axis around which a control surface moves.

horizontal stabilizer The entire horizontal part of an airplane's empennage comprising both fixed and movable surfaces. On most airplanes, the horizontal stabilizer is the greatest contributor to longitudinal stability.

horsepower A unit of power equal to 550 ft-lbs/sec or 33,000 ft-lbs/min.

humidity The amount of water vapor in the air.

hypersonic Movement or flow at very high supersonic speeds, generally at a Mach number of 5 or greater.

indicated airspeed (IAS) The instrument indication for the amount of dynamic pressure that the aircraft is creating during some given flight condition. Indicated airspeed is displayed in knots, abbreviated KIAS.

indicated altitude The indication on a pressure altimeter when the kollsman window is set to the current local altimeter setting.

induced drag (D$_I$) That portion of total drag resulting from the production of lift.

infinite wing A wing with no wingtips; used in discussing airflow around an airfoil in ideal situations.

interference drag Drag caused by the mixing of streamlines around aircraft components due to their proximity. It is a form of parasite drag.

inverted spin A spin characterized by negative g's and an inverted attitude.

isothermal layer The layer of the atmosphere from approximately 36,000 through 66,000 feet, in which the air remains at a constant temperature of $-56.5°C$.

kinetic energy (KE) The ability of a body to do work because of its motion.

laminar flow The portion of the boundary layer airflow that is smooth and unbroken and travels along well defined streamlines.

laminar flow wing An airfoil specially designed to maintain a laminar flow boundary layer.

lateral axis An axis going through an airplane's center of gravity from side to side (wingtip to wingtip). Any movement developed around this axis is called pitch.

lateral control Control of the lateral axis around the longitudinal axis; roll control.

lateral moment A moment created around an airplane's longitudinal axis.

lateral stability The stability of an aircraft around its longitudinal axis. The reaction of an aircraft to an angle of bank.

leading edge flaps A high lift device consisting of a hinged portion of the leading edge of a wing that moves down to increase the wing's camber.

leading edge radius The radius of a circle tangent to the leading edge, upper and lower surfaces of the airfoil.

lift (L) The component of the aerodynamic force acting perpendicular to the relative wind.

lift to drag ratio (L/D) The ratio of lift to drag, obtained by dividing the coefficient of lift by the coefficient of drag. A measure of the wing's efficiency. The L/D ratio is also used as the glide ratio.

lift to drag ratio-maximum (L/D$_{MAX}$) The greatest ratio of lift to drag. L/D$_{MAX}$ AOA is the most efficient AOA for that airfoil.

limit airspeed See redline airspeed.

limit load The maximum load factor an airplane can sustain without any possibility of permanent deformation. It is the maximum load factor anticipated in the normal operation of the airplane.

linear acceleration Acceleration along a line or axis.

load A stress-producing force.

load factor (n) The ratio of the load applied by an airplane's lift to the load applied by its weight. It is a multiple of the acceleration of gravity, commonly called "Gs."

local speed of sound The speed at which sound travels in a given medium under local ambient conditions.

longitudinal axis An axis extending from the nose to the tail of an aircraft, passing through its center of gravity. Any movement developed around this axis is called roll.

longitudinal control Control of the longitudinal axis around the lateral axis; pitch control.

longitudinal moment A moment created around an airplane's lateral axis.

longitudinal stability The stability of an aircraft around the lateral axis. The reaction of an aircraft to changes in pitch.

Mach number (M) (Pronounced "mock," after Ernest Mach (1838-1916), Austrian scientist.) The ratio of the true airspeed of an object moving through the air to the local speed of sound in that air.

maneuverability The ability of an airplane to readily alter its flightpath. The ease with which an airplane moves out of equilibrium.

maneuver point The point on the V-n diagram at the intersection of the positive accelerated stall line and the positive limit load. It is the point where the limit load may be achieved without the possibility of overstress, or the lowest airspeed that the limit load is encountered.

maneuver speed (V_a) The indicated airspeed that an airplane can achieve its maximum turn rate and minimum turn radius. The slowest velocity that an airplane can generate its limit load. It is usually the recommended turbulent air penetration airspeed.

mass (m) The quantity of molecular material that comprises an object.

mass balance The feature of a control surface that reduces the magnitude of the inertial and gravitational moments around the hinge line.

mean aerodynamic chord (MAC) The chord of an imaginary rectangular airfoil that would have pitching moments throughout the flight range the same as those of an actual airfoil.

mean camber line A line halfway between the upper and lower surface of an airfoil.

minimum glide angle The smallest angle between the horizon and the flightpath of an airplane in a glide.

moment A tendency to cause rotation around a point or axis, as a control surface around its hinge or an airplane around its center of gravity; the measure of this tendency, equal to the product of the force and perpendicular distance between the point of rotation and the direction of the force., expressed as a vector. Also called torque.

moment arm The distance from a point of rotation, perpendicular to the force, over which a force acts to create a moment.

monocoque A type of construction, as an airplane fuselage, in which most or all the stresses are carried by the covering or skin.

nacelle A streamlined structure or compartment on an aircraft, used as housing for an engine.

negative camber airfoil An airfoil in which the mean camber line is below the chordline.

neutral point (NP) The location of the center of gravity of an airplane that would produce neutral longitudinal static stability. The average aerodynamic center for the overall airplane.

Newton's First Law (The Law of Equilibrium.) "A body at rest tends to remain at rest and a body in motion tends to remain in motion in a straight line at a constant velocity unless acted upon by some unbalanced force."

Newton's Second Law (The Law of Acceleration.) "The acceleration of a body is directly proportional to the force exerted on the body, is inversely proportional to the mass of the body, and is in the same direction as the force." $F = m\,a$.

Newton's Third Law (The Law of Interaction.) "For every action, there is an equal and opposite reaction."

nosewheel liftoff / touchdown speed (NWLO/TD) The lowest speed that a heading and course along the runway can be maintained with full rudder and ailerons deflected when the nosewheel is off the runway.

overstress The condition of possible permanent deformation or damage that results from exceeding the limit load. It also refers to the damage that may occur as a result of exceeding the limit load. Overstress damage will not cause structural failure of the airframe, but could result in internal damage to various components.

parasite drag (D_P) All drag not associated with the production of lift.

phugoid oscillations Oscillations of altitude and airspeed that occur over relatively long periods of time, and are easily controlled by the pilot. Also called phugoid motion.

pilot induced oscillations (PIO) Oscillations of attitude and angle of attack caused by the pilot trying to stop unwanted aircraft oscillations, or by the instability of the control surfaces. These inputs may result in an increase in the magnitude of the original oscillations.

pitch The motion of an aircraft around its lateral axis. Pitch control is achieved through use of elevators or stabilators.

pitch angle The angle between the chordline of the rotor blade and the rotor's tip path plane; the angle between the propeller blade and the propeller tip path plane.

pitch attitude (θ) The angle between the longitudinal axis of the airplane and the horizon.

pitching moment Any moment around the lateral axis of an airplane.

pitot static system A system consisting of a pitot tube, a static pressure port, and a device that determines the difference, used principally in order to calculate dynamic pressure.

plain flap A high lift device consisting of a hinged airfoil attached to the leading or trailing edge of a wing that increases camber to increase the C_L.

planform The outline of an object, such as a wing, as viewed from above.

positive camber airfoil An airfoil in which the mean camber line is above the chordline.

potential energy (PE) The ability of a body to do work because of its position or physical state.

power (P) The rate of doing work, or work per unit time, measured in ft-lbs/sec or horsepower.

power available (P_A) The power an engine is producing. Power available is a function of PCL setting, density altitude, and velocity.

power control lever (PCL) Control on a propeller driven airplane or helicopter, that adjusts the fuel flow and therefore the power output of the engine(s). Similar to the throttle on a jet aircraft.

power deficit (P_D) The negative difference between power available and power required.

power excess (P_E) The positive difference between power available and power required.

power required (P_R) The power required to produce enough thrust to overcome drag in level equilibrium flight.

pressure altimeter Aneroid barometers calibrated to indicate altitude in feet instead of pressure.

pressure altitude (PA) Height above the standard datum plane, i.e., altitude measured from standard sea level pressure by a barometric altimeter.

pressure gradient A change in the pressure of a fluid per unit of distance.

propeller efficiency (p.e.) A measure of the effectiveness of a propeller in converting shaft horsepower into thrust horsepower.

propeller wash The disturbed air produced by the passage of the propeller, usually making a corkscrew path around the airplane.

pullout An act or instance of recovering from a dive.

radar altimeter Specialized radar transmitter/receiver used to indicate height above terrain.

radius of turn (r) See turn radius.

range The distance that an aircraft can travel without refueling.

rate of climb (ROC) The rate at which an aircraft gains altitude, the vertical component of its airspeed in a climb.

rate of descent (ROD) The rate at which an aircraft loses altitude, the vertical component of its airspeed in a descent. Also called sink rate.

rate of turn (ω) See turn rate.

redline airspeed (V_{NE}) The maximum permissible airspeed for an airplane. Beyond the redline airspeed, a pilot may experience control problems and structural damage to the aircraft due to aeroelastic effects.

region of normal command The region of flight at velocities greater than maximum endurance airspeed in which an airplane is in stable equilibrium. That is, if disturbed (slowed down), it tends to return to equilibrium.

region of reversed command The region of flight at velocities less than maximum endurance airspeed, in which a greater power setting is required to fly at a lower velocity, due to increased total drag caused by induced drag. Takeoff and landing normally take place while in this region. Also called the "back side of the power curve."

relative wind (RW) The airflow experienced by the aircraft as it flies through the air. It is always equal and opposite to the flightpath. The relative wind may arise from the motion of the body, from the motion of the air, or from both.

reverse thrust Thrust applied to a moving object in a direction opposite to the direction of the object's motion.

reversibility The ability to transmit aerodynamic forces from the control surfaces to the cockpit controls.

roll The motion of an airplane around its longitudinal axis. Roll is controlled by the use of ailerons or spoilers.

rudder An upright control surface that is deflected to produce a yawing moment, rotating the airplane around its vertical axis.

safe flight envelope The portion of the V-n diagram that is bounded on the left by the accelerated stall lines, on the top and bottom by the positive and negative limit loads, and on the right by redline airspeed. An aircraft my operate in its safe flight envelope without exceeding its structural or aerodynamic limits.

scalar A quantity expressing only magnitude, e.g., time, amount of money, volume of a body.

section A cross section of an airfoil taken at right angles to the span axis or some other specified axis of the airfoil.

semi-monocoque A type of construction, as of a fuselage or nacelle, in which transverse members and stringers reinforce the skin and help carry the stresses.

shaft horsepower The horsepower delivered at the rotating driveshaft of an engine.

shielded horn The part of a control surface of longer chord than the rest of the surface, lying forward of the hingeline and partially shielded by the surface to which it is attached, used for aerodynamic balance.

shockwave A surface or sheet of discontinuity set up in a supersonic field of flow, through which the fluid undergoes a finite decrease in velocity accompanied by a marked increase in pressure, density, temperature, and energy.

sideslip A movement of an airplane such that the relative wind has a component parallel to the lateral axis.

sideslip angle (β) The angle between the airplane's longitudinal axis and the relative wind, as seen from above.

sideslip relative wind The component of the relative wind that is parallel to the airplane's lateral axis.

sine (sin) In a right triangle, the function of an acute angle that is the ratio of the length of the opposite side to the length of the hypotenuse.

sink rate See rate of descent.

skin friction The friction of a fluid against the skin of an aircraft or other body; friction drag.

slat The vane used in a slot, especially an automatic slot. When the slat deploys it forms a slot.

slot High lift device that consists of a fixed vane that forms a gap between the leading edge of the wing and the body of the wing that allows airflow from below the wing to reach the upper surface and reenergize the boundary layer, delaying boundary layer separation. Also called fixed slot.

slotted flap A high lift device consisting of a hinged airfoil attached to the leading or trailing edge of a wing that increases camber and uses BLC to increase the C_L.

sonic Pertaining to sound or the speed of sound.

sonic boom An explosion-like sound heard when a shock wave generated by a supersonic airplane reaches the ear.

sonic speed Speed equal to the speed of sound.

sound barrier A popular term for the large increase in drag that acts upon an aircraft approaching the speed of sound.

span See wingspan.

spanwise flow Airflow that travels the span of the wing, parallel to the leading edge, normally root to tip. This airflow is not accelerated over the wing and therefore produces no pressure differential or lift.

spar A principal spanwise beam in the structure of a wing.

speed of sound The speed at which sound travels in a given medium under certain conditions. The speed of sound in air is primarily dependent on the temperature of the air mass.

spin An asymmetrical aggravated stall resulting in autorotation.

spiral A maneuver in which an airplane ascends or descends in a helical (corkscrew) path at an angle of attack within the normal range of flight angles.

spiral divergence A motion resembling a spiraling descent, becoming steeper over time. Spiral divergence results from strong static directional stability and weak static lateral stability.

split flap A high lift device consisting of a plate deflected from the lower surface of the trailing edge of a wing that increases camber to increase the C_L. It produces a similar change in C_L as a plain flap, but a much larger increase in drag due to the great turbulent wake produced.

spoiler A movable control surface attached to the wing of an airplane, used to produce a rolling moment around the longitudinal axis by disturbing the flow of the boundary layer over one wing.

stabilator A movable control surface that replaces the horizontal stabilizer and elevators.

stability The property of a body, such as an aircraft, to maintain its attitude or to resist displacement, and if displaced, to develop forces and moments that would return it to its original condition.

stabilizer A fixed or adjustable airfoil or vane that provides stability for an aircraft, i.e., a fin, the horizontal or vertical stabilizer on an airplane.

stagnation Loss of kinetic energy or velocity. Lack of motion.

stall A condition of flight in which an increase in AOA will result in a decrease in C_L.

stall fence A plate or vane projecting from the upper surface of a wing, parallel to the airstream, used to prevent spanwise flow.

stalling angle of attack The angle of attack on an airfoil beyond which a stall occurs, i.e., $C_{L_{MAX}}$. Beyond this angle of attack, the boundary layer is unable to remain attached to the wing, resulting in the decrease in C_L.

stall speed (V_S) The minimum true airspeed required to maintain level flight at $C_{L_{MAX}}$ AOA.

stall strip A sharply angled device attached near the wing's root on its leading edge to initiate a root first stall.

standard atmosphere A reference set of average atmospheric conditions.

standard datum plane (SDP) The actual elevation at which the barometric pressure is 29.92 inHg.

standard rate turn (SRT) A turn in an aircraft with a three degree per second turn rate.

static failure The breaking (or serious permanent deformation) of a material due to a steadily increasing, or sudden large application of force. This type of failure is often immediate and can occur without warning.

static pressure (P_S) The weight of a column of air over a given area; the pressure each air particle exerts on another due to the weight of all the particles above; the potential energy per unit volume.

static stability The initial tendency of an object to either move toward or away from equilibrium after a disturbance.

static strength A measure of a material's resistance to a steadily increasing load or force.

steady airflow Airflow in which at every point in the moving air mass, the pressure, density, temperature and velocity are constant.

stiffness Resistance to deflection or deformation.

straight horn See unshielded horn.

streamline The path traced by a particle of air while in steady flow.

streamtube An impenetrable tube formed by many streamlines. Streamtubes are closed systems.

strength A measure of a material's resistance to load or force.

subsonic Movement or flow at speeds below the speed of sound, generally at a Mach number of 0.0 to 0.75.

supersonic Movement or flow at speeds above the speed of sound, generally at a Mach number of 1.2 to 5.0.

sweep angle (Λ) The angle measured between the line of 25% chord and a line drawn perpendicular to the root chord. Also called sweepback.

symmetric Exhibiting a correspondence of parts on opposite sides of a boundary or axis

symmetric airfoil An airfoil in which the mean camber line is coincident with the chordline. Also called a zero camber airfoil.

tangent (tan) In a right triangle, the function of an acute angle that is the ratio of the length of the opposite side to the length of the adjacent side. A line, curve, or surface touching but not intersecting another line, curve or surface at only one point.

taper A gradual reduction in the chord length of an airfoil from root to tip.

taper ratio (λ) The ratio of tip chord to root chord. The taper ratio affects the lift distribution and the structural weight of the wing.

temperature A measure of the average kinetic energy of air particles, expressed in degrees Celsius (°C), Fahrenheit (°F), or Kelvin (K).

terminal velocity The maximum velocity an airplane can attain under given conditions. A vertical (zero-lift) dive path, normal gross weight, zero engine thrust, and standard sea-level air density are assumed.

thickness The cross sectional height of an airfoil measured perpendicular to the chordline.

thrust available (T_A) The thrust an engine produces under a specific velocity, density, and throttle setting.

thrust axis The axis along which thrust is produced and the direction in which the force is generated.

thrust deficit (T_D) The negative difference between thrust available and thrust required.

thrust excess (T_E) The positive difference between thrust available and thrust required.

thrust horsepower The actual amount of horsepower that an engine-propeller system transforms into thrust, equal to shaft horsepower multiplied by propeller efficiency.

thrust required (T_R) The thrust required to overcome drag to maintain level equilibrium flight.

total pressure The pressure a moving fluid would have if it were brought to a rest without losses.

transonic Movement or flow at speeds very near the speed of sound, generally at a Mach number of 0.75 to 1.2.

trimmed flight A condition that exists when the sum of the moments acting around the center of gravity are equal to zero. The word "trim" often refers to the balance of control forces.

trim tab A tab that is deflected to a position where it remains to keep the aircraft in the desired attitude.

true airspeed (TAS) The velocity of an aircraft with respect to the air mass in which it is traveling. Airspeed value determined by correcting indicated airspeed for installation error, compressibility, and density.

true altitude The actual height above mean sea level. It is found by correcting calibrated altitude for temperature deviations from the standard atmosphere.

turbulence An agitated condition of air in which random fluctuations in velocity and direction occur. Airflow in which the velocity at any point varies erratically in magnitude and direction.

turbulent flow Boundary layer airflow characterized by turbulent, unsteady airflow.

turn radius (r) One half the diameter of the circle an aircraft would fly if it completed a 360 degree turn.

turn rate (ω) The number of degrees of arc traversed per unit of time while turning, expressed in degrees/sec.

ultimate load The maximum load factor that the airplane can withstand without structural failure. It is 1.5 times the limit load.

unshielded horn The part of a control surface of longer chord than the rest of the surface, lying forward of the hingeline and entirely exposed to the relative wind, used for aerodynamic balance.

upwash Chordwise airflow from the lower surface of an airfoil passing upward over the leading edge to the upper surface. Any upward airflow.

V-n diagram A diagram describing the structural and aerodynamic limits within which an airplane must operate.

vector A quantity that expresses both magnitude and direction. A vector quantity is represented by an arrow that displays direction and has a length proportional to magnitude.

velocity (V) Speed, as referenced to another plane, object, or system. A vector quantity equal to speed in a given direction. True airspeed.

velocity never-to-exceed (V_{NE}) See redline airspeed.

vertical axis An axis passing from top to bottom through the aircraft's center of gravity. Any movement developed around this axis is called yaw.

vertical stabilizer A fin mounted approximately parallel to the plane of symmetry of an airplane, to which the rudder is attached.

viscosity (μ) A measure of a fluid's resistance to flow and shearing.

volume The size of the mass, or the amount of space occupied by an object.

vortices / wingtip vortices A spiraling mass of air created at the wingtip, due to the airflow around the tip from the high-pressure region below the surface to the low-pressure region above it. Vortex strength is dependent upon the wing loading, gross weight, and speed of the generating airplane. Vortices from medium to heavy airplanes can be extremely hazardous to smaller airplanes. Also called wake turbulence, or jetwash.

weight The force at which a mass is attracted toward the center of the earth by gravity.

wing An airfoil that produces a pressure differential when air is forced over it, resulting in a lifting force.

wing area (S) The surface area of a wing from wingtip to wingtip. The area within the outline of a projection of a wing on the plane of its chord, including that area lying within the fuselage or nacelles. With a swept wing, the area within the fuselage is contained within lines having the same sweep angle as the leading and trailing edges, fairings or fillets being ignored.

wing loading (WL) A ratio of airplane weight to the wing surface area.

wing root The base of a wing, where it joins the fuselage or other main body of an airplane.

wing section A cross section of a wing; the profile of a cross section or the area defined by a profile.

wingspan (b) The length of a wing, measured from wingtip to wingtip. Also called span.

work (W) Work is done when a force acts on a body and it moves. Work is a scalar quantity measured in ft-lbs. $W = F \times s$

yaw Rotation around the vertical axis of an airplane. Yaw is controlled by the rudder.

Answers to Study Questions

TOPIC 1:

1.   A vector quantity expresses both magnitude and direction, while a scalar quantity expresses only magnitude.

2.   Mass is the quantity of molecular material that comprise an object.

3.   Weight is the force at which a mass is attracted toward the center of the earth by gravity.

4.   The density of the air is the mass of air per unit of volume.

5.   A force is a push or pull exerted on a body. It tends to produce motion along a line. A moment is a tendency to produce motion about a point or axis. It is created by applying force to a lever arm to induce rotational motion about an axis.

6.   Work is done when a force acts on a body and moves it.

$$W = F \cdot s$$

7.   Power is the rate of doing work.

8.   Energy is the ability to do work.

$$TE = PE + KE$$

9.   Potential energy is the ability of a body to do work because of its position or state of being.

10.   Kinetic energy is the ability of a body to do work because of its motion.

$$KE = \tfrac{1}{2}mV^2$$

11.   Newton's First Law of Motion is the Law of Equilibrium: "A body at rest tends to remain at rest and a body in motion tends to remain in motion in a straight line at a constant velocity unless acted upon by some unbalanced force."

12.   An airplane traveling at a constant speed and direction, and an airplane parked on the flight line are in equilibrium if the sum of the forces and moments about the center of gravity equal zero.

13.   Trimmed flight exists when the sum of the moments acting about the C.G. equals zero, where equilibrium flight exists when the sum of the forces <u>and</u> moments about the C.G. equal zero.

14.   Newton's Second Law of Motion is the Law of Acceleration: "An unbalanced force (F) acting on a body produces an acceleration (a) in the direction of the force that is directly proportional to the force and inversely proportional to the mass (m) of the body." An example is an aircraft in a turn.

15.   Newton's Third Law of Motion is the Law of Interaction: "For every action there is an equal and opposite reaction."

16.   Static pressure is the weight of a column of air over a given area. Static pressure decreases with an increase in altitude.

17.   Air density decreases with an increase in altitude.

18.   Air temperature is a measure of the average kinetic energy of air particles.

19.   Air temperature decreases by 2.0°C per 1,000 feet until approximately 36,000 feet. An isothermal layer with a constant temperature of about −56.5°C exists from approximately 36,000 feet through 66,000 feet.

20.   Air density decreases with an increase in humidity.

21.   Air viscosity is a measure of air's resistance to flow and shearing. Air viscosity increases with an increase in temperature.

22.   The primary factor in determining the speed of sound in air is temperature.

23.

	English	Metric (SI)
Static Pressure P_{S0}	29.92 inHg	1013.25 mbar
Temperature T_0	59 °F	15 °C
Average Lapse Rate	3.57 °F / 1000 ft	2 °C / 1000 ft
ρ_0	.0024 slugs / ft^3	1.225 g / l
Local Speed of Sound	661.7 knots	340.4 m / s

24. $P = \rho RT$. Given a constant pressure, density will decrease with an increase in temperature.

TOPIC 2:

1.   A heavier-than-air fixed-wing aircraft that is driven by an engine-driven-propeller or a gas turbine jet and is supported by the dynamic reaction of airflow over its wings.

2.   The T-34C uses a semi monocoque fuselage consisting of stringers, transverse frame members, and the skin.

3.   Full cantilever

4.   The ailerons are attached to the wing.

5.   The rudder and elevator are attached to the empennage.

6.   The elevator is used for longitudinal control.

7.   The rudder is the primary source of directional control.

8.   The center of gravity is the point at which all weight is considered to be concentrated.

9.   Longitudinal-roll; lateral-pitch; directional-yaw.

10.   Wingspan is the length of a wing, measured from wingtip to wingtip.

11.   The chordline is an imaginary straight line drawn through the leading and trailing edges of an airfoil. Chord is the length of the chordline from the leading edge to the trailing edge of the airfoil. Average chord is the geometric average of every chord from the root to the tip. Tip chord is the length of the chord at the wing tip. Root chord is the length of the chord at the wing root.

12. The apparent surface area of the wing, from wingtip to wingtip, including the area within the fuselage and nacelles.

$$S = bc$$

13. Taper is the reduction in the chord from wing root to tip. Taper ratio (λ) is the ratio of the tip chord to the root chord.

$$\lambda = \frac{c_T}{c_R}$$

Sweep angle is the angle between a line drawn 25% aft of the leading edge and a line parallel to the lateral axis.

14. The ratio of the wing span to the average chord. A B-52 bomber would have a high aspect ratio. A high performance fighter would have a low aspect ratio.

15. The angle formed between the chordline of an airfoil and the longitudinal axis of the airplane. It is fixed on most airplanes.

16. The ratio of an airplane's weight to the surface area of it's wings.

$$WL = \frac{W}{S}$$

17. The angle between the spanwise inclination of the wing and the lateral axis.

TOPIC 3:

1. $\rho_1 A_1 V_1 = \rho_2 A_2 V_2$. ρ is density, A is cross sectional area, and V is velocity. ρ may be cancelled if altitude remains constant and airflow is subsonic.

2. reduced by one-half.

3. $P_T = P_S + q$. Total pressure remains constant in a closed system. P_S is inversely related to q, if P_T is constant.

4. The pitot-static system consists of a total pressure sensor, the pitot tube; a static pressure sensor, the static port; and a mechanism that determines the difference between the two in order to calculate dynamic pressure which is displayed in the cockpit as indicated airspeed ($q = P_T - P_S$).

5. For a given altitude, the pressure in the static pressure port of the airspeed indicator is constant for all airspeeds and all angles of attack.

6. IAS is the instrument indication for the dynamic pressure the aircraft is creating during flight. True airspeed (TAS) is the actual velocity at which an aircraft moves though an air mass.

$$TAS = \sqrt{\frac{\rho_0}{\rho}} IAS$$

7. IAS will equal TAS when $\rho = \rho_0$. As altitude increases with a constant IAS, TAS will increase.

8. Decrease the indicated airspeed as altitude increases.

9. 260 TAS.

10. 120 knot headwind.

11. Mach number is a ratio of the airplane's true airspeed (TAS) to the local speed of sound (LSOS). Critical Mach number is the free airstream Mach number (<1) that produces the first evidence of local sonic flow.

12. Mach Number increases because TAS increases and the local speed of sound decreases.

TOPIC 4:

1. D. The angle between the longitudinal axis and the horizon.

2. Flight path is the path described by the airplane's center of gravity as it moves through the air mass.

3. Relative wind is the apparent wind created by the airplane's movement through the air. It is equal in magnitude and opposite in direction to the flight path.

4. C. Between the relative wind and the chordline.

5. Mean camber line is a line drawn halfway between the upper and lower surfaces of the wing. The mean camber line is above the chordline in a positive camber airfoil, below the chordline in a negative camber airfoil, and is coincident with the chordline in a symmetric airfoil.

6. The aerodynamic center is the point along the chordline where all changes in the aerodynamic force take place.

7. On the upper surface of the airfoil, dynamic pressure increases (due to an increase in airflow velocity), and static pressure decreases. A static pressure differential is created between the upper and lower surfaces which produces a force perpendicular to the relative wind.

8. Aerodynamic force is a force that is the result of pressure and friction distribution over an airfoil.

$$AF = \tfrac{1}{2}\rho V^2 S C_F$$

9. Lift and drag.

10. The pilot can normally control velocity, camber, and angle of attack.

$$L = \tfrac{1}{2}\rho V^2 S C_L$$

11. Increase angle of attack.

12. A symmetric airfoil produces zero lift at zero α, while a cambered airfoil must be taken to some negative α to produce zero lift. Also, C_{Lmax} for a cambered airfoil is higher but stall occurs at a lower α than on a symmetric airfoil.

13. C_{Lmax} AOA

14. The lift vector is always perpendicular to the relative wind.

15 Lift will increase and the airplane will climb.

16. Lift will decrease and the airplane will descend

17. If the changes in AOA and airspeed are properly coordinated, the airplane will speed up and maintain level flight.

18. The airplane will stall.

19. Lift will decrease and the airplane will descend.

20. Laminar flow is smooth, unbroken airflow that travels along well defined streamlines. Turbulent flow is disorganized and the streamlines break up. Laminar flow produces little friction but is easily separated from the airfoil. Turbulent flow produces more friction, but does not separate from the airfoil as easily.

21. The primary feature of airflow separation is stagnation of the lower levels of the boundary layer. During boundary layer separation, the separation point moves forward along the airfoil.

22. A stall is a condition of flight in which an increase in AOA results in a decrease in C_L. A stall is caused by exceeding C_{Lmax} AOA.

23. C. Has no effect on stalling angle of attack

24. D. C_L decreases and lift decreases.

25. Stall speed is the minimum true airspeed required to maintain level flight at C_{Lmax} AOA.

$$V_s = \sqrt{\frac{2W}{\rho S C_{Lmax}}}$$

26. True stall speed increases with an increase in weight or altitude. Indicated stall speed increases with an increase in weight, but remains constant with an increase in altitude.

27. C. power on, 9

28. High lift devices decrease stall speed by increasing C_{Lmax}. Camber change devices decrease stalling AOA, and BLC devices increase stalling AOA.

29. Slots and slats. Slots take high pressure air from beneath the wing, increase its velocity through a nozzle and inject it into the lower levels of the boundary layer on the upper surface of the airfoil.

30. A. Increases lift and increases drag

31. Plain flaps, split flaps, slotted flaps, fowler flaps, leading edge plain flaps, and leading edge slotted flaps. The T-34C has slotted flaps. Fowler flaps produce the greatest increase in C_{Lmax}.

32. To maintain aileron effectiveness in the early stages of stall, and provide an aerodynamic stall warning.

33. Geometric twist decreases the angle of incidence from root to tip, and aerodynamic twist changes the airfoil shape (camber and/or relative thickness) from root to tip.

34. Simultaneously add power, relax back stick pressure, and roll wings level.

TOPIC 5:

1. Drag is the component of aerodynamic force that is parallel to the relative wind and acts in the same direction. ρ, V, S, camber, μ, α, AR, compressibility.

2. The coefficient of drag increases with an increase in angle of attack. Never.

3. Parasite drag is all drag not associated with the production of lift. Parasite drag consists of form drag, friction drag, and interference drag.

4. D_P = q f. q is dynamic pressure; f is the equivalent parasite area.

5. Form drag may be reduced by streamlining. Friction drag may be reduced by cleaning and polishing. Interference drag may be reduced by proper filleting and fairings.

6. Parasite drag increases as velocity increases.

7. High pressure air in the leading edge stagnation point flows up and around the leading edge of the airfoil, creating "upwash". Some airflow also moves down around the trailing edge, and is called "downwash". On an infinite wing, upwash equals downwash with no net effect on lift.

8. Airflow on a finite airfoil develops a spanwise component toward the wingtip. This spanwise airflow passes around the wingtip to the top of the wing, and causes the downwash to be approximately double the upwash. This increased downwash is responsible for decreasing effective lift, increasing drag, and forming wingtip vortices.

9. Induced drag is the portion of total drag associated with the production of lift. It is formed on finite wings because the airflow moving around the wingtip increases downwash. Downwash causes the lift vector to rotate aft. The component of this vector that is parallel to the free airstream relative wind is induced drag.

10. L is lift, V is velocity, ρ is density, b is wingspan, and k is a constant.

$$D_I = \frac{kL^2}{\rho V^2 b^2}$$

11. Induced drag may be reduced by increasing velocity, wingtip devices, or by increasing wingspan.

12. Induced drag varies inversely with the square of velocity, and therefore decreases as velocity increases.

13. $D = \frac{1}{2}\rho V^2 S C_D$ $D_T = D_P + D_I$

14. B, C

15. The lift to drag ratio is a measure of the wing's efficiency. It is calculated using the ratio of the coefficients of lift and drag. Since both C_L and C_D depend on angle of attack, the L/D ratio will be determined by the angle of attack at which the airplane is flying.

16. L/D_{MAX} is the greatest ratio of lift to drag. It is easily recognizable as the lowest point on the total drag curve. L/D_{MAX} is the most efficient AOA, the point of minimum total drag, the AOA where $D_P = D_I$, and where the wing produces its greatest L/D ratio.

TOPIC 6:

1.   The amount of thrust required to overcome drag and maintain equilibrium flight. The drag curve.

2. $P_R = \dfrac{T_R \cdot V}{325}$

3.   C. at, slower than

4. T_A (or P_A) is the amount of thrust (or power) the airplane's engine is currently producing at a given throttle setting, velocity and air density.

5.   PT6A-25, 550, 425

6.   C. equal to L/D_{MAX} AOA, greater than L/D_{MAX} AOA

7.   D. greater than L/D_{MAX}, equal to L/D_{MAX}

8.   The T_R and P_R curves move up and to the right with a weight increase. The T_R curve shifts to the right and the P_R curve shifts up and to the right with an altitude increase.

9. T_A and P_A increase as throttle/PCL setting increases, decrease as altitude increases, and are unaffected by weight.

10.   An increase in altitude or weight causes a decrease in P_E and T_E.

TOPIC 7:

1.   The power curves represent fuel flow for a turboprop; thrust curves represent fuel flow for a turbojet.

2.   Maximum endurance is the maximum time airborne for a given amount of fuel. Maximum range is the maximum distance over the ground for a given amount of fuel.

3.   C. Prop maximum range, jet maximum endurance

4.   Maximum endurance and maximum range performance will decrease. Both airspeeds will increase.

5.   The higher TAS along with better fuel efficiency will yield an increase in range and endurance.

6.   A tailwind will increase max range but will not affect endurance. Max range airspeed will decrease, max endurance airspeed will not be affected.

7.   Fuel flow is greatly increased.

8.   Maximum angle of climb is the angle that achieves the greatest altitude for the minimum distance covered over the ground.

9.   D. Full throttle, jets at L/D_{MAX}, props slower than L/D_{MAX}

10.   B. Maximum rate of climb angle of attack is smaller than max angle of climb.

11.   Maximum angle of climb, 75 KIAS, 125 KIAS.

12.   An increase in altitude or weight will decrease the excess thrust and power resulting in a decrease in maximum rate of climb and maximum angle of climb.

13.   Absolute ceiling

14.   At an AOA greater than L/D_{MAX} and a velocity less than L/D_{MAX}.

15.   At L/D_{MAX} AOA and velocity.

16.   An increase in altitude will increase both glide range and glide endurance. An increase in weight decreases glide endurance but does not affect glide range. A headwind will decrease glide range and have no effect on glide endurance.

17.   D. 20 to 22 nautical miles

18.   Windmilling propeller

19.   A. Decrease in airspeed requires an increase in throttle

20.   If an airplane increases angle of attack without increasing the throttle, it will develop a deficit which causes either a deceleration or a descent.

TOPIC 8:

1.   Ailerons and spoilers are used for roll control. The rudder is used to yaw. The elevators are used for pitch control only.

2.   A nose-up pitch is created by moving the stick aft to move the elevator upward.

3.   A right roll is created by moving the stick right to move the left aileron down and the right aileron up.

4.   Trimming reduces the force required to hold control surfaces in a position necessary to maintain a desired flight attitude. The elevator trim tab must be moved down to hold the elevator up, causing a nose-up pitch attitude.

5.   A shielded horn provides aerodynamic balance for the rudder and elevator.

6.   An overhang provides aerodynamic balance for the T-34C aileron.

7.   A lead weight in the shielded horn provides mass balance for the T-34C elevator.

8.   They use conventional controls. In the conventional type control system the stick and rudder pedals are directly connected to the control surfaces via push-pull tubes, pulleys, cables and levers.

9.   The aileron trim tab is servo, the elevator is neutral, and the rudder has an anti-servo type trim tab. The servo type helps the pilot to deflect the aileron.

10.   The elevator uses a combination of bobweights and a downspring to provide the pilot with some artificial feel.

TOPIC 9:

1.   Static stability is the initial tendency of an object to move toward or away from its original equilibrium position. Dynamic stability is the tendency of an object to return toward or move away from equilibrium, with respect to time.

2.	Divergent oscillation is associated with positive static and negative dynamic stability.

3.	Maneuverability and stability are inversely related.

4.	Longitudinal stability concerns the stability of the longitudinal axis around the lateral axis. This is the motion of pitch.

5.	The horizontal stabilizer is the greatest positive contributor to longitudinal static stability. The fuselage and straight wings are negative contributors to longitudinal static stability. Wing sweep is a positive contributor to longitudinal static stability. If the C.G. is forward of the neutral point, the overall airplane has positive longitudinal stability.

6.	Sweeping the wing aft moves the A.C. aft.

7.	Surface area and distance from the C.G.

8.	The angle between the longitudinal axis and the relative wind.

9.	Directional stability concerns the stability of the longitudinal axis around the vertical axis. This motion is yaw.

10.	The fuselage is a negative contributor to directional static stability. The vertical stabilizer is the greatest positive contributor to directional stability. Straight and swept wings are positive contributors to directional static stability.

11.	Lateral stability concerns the stability of the lateral axis around the longitudinal axis. This motion is roll.

12.	Wing dihedral is the greatest positive contributor to lateral static stability. A low mounted wing is laterally destabilizing and a high mounted wing will be laterally stabilizing. Swept wings, dihedral, and the vertical stabilizer are laterally stabilizing. Anhedral is laterally destabilizing.

13.	Directional divergence is a result of negative directional static stability. Spiral divergence results from strong directional and weak lateral static stability. Dutch roll is the result of strong lateral and weak directional stability.

14.	Adverse yaw. The additional drag on the up-going wing due to increased camber.

15.	Proverse roll. Increased velocity on the outer wing creates more lift.

16.	Opposite rudder must be used.

17.	If the relative wind is below the thrust line, the down-going blade will create more thrust and will yaw the nose to the left.

18.	left, torque, left.

TOPIC 10:

1.	A spin is an aggravated stall that results in autorotation. Stall and yaw.

2.	The up-going wing has a lower AOA, more lift and less drag than the down-going wing, which has a higher AOA and therefore more drag and less lift.

3.	AOA, airspeed and turn needle (the altimeter is monitored to ensure compliance with bailout/ejection criteria).

4.   A heavier aircraft will have a slower spin entry due to a greater inertial moment.

5.   The lower stall speed will result in a slower spin entry.

6.   C. Break the stall and stop rotation.

7.   An aggravated spin.

8.   Positive-g stall entry with yaw.

9.   Right.

10.   Recover from unusual attitude.

11.   D. Direction of spin always.

TOPIC 11:

1.   The total lift vector must increase in magnitude.

2.   The vertical component of lift must equal airplane weight. The horizontal component turns the airplane.

3.   Load factor or Gs.

4.   B. Angle of bank

5.   B. 2.9

6.   As the AOB approaches 90°, the load factor approaches infinity. An airplane can maintain 90° AOB only by creating lift on a source other than the wings, such as the fuselage, vertical stabilizer, etc.

7.   Stall speed increases during turning or maneuvering flight because of the increase in the load factor.

8.   Static strength is a resistance to a single application of force. Fatigue strength is a resistance to a cyclic application of force.

9.   Static failure is the breaking of a material due to a single application of a steadily increasing load or force. Fatigue failure is the breaking of a material due to a cyclic application of load or force.

10.   Limit load factor is the greatest load factor an airplane can sustain without any risk of permanent deformation. Exceeding the limit load factor will cause overstress.

11.   +4.5 Gs, −2.3 Gs

12.   elastic limit, elastic limit

13.   Ultimate load factor is the maximum load factor that the airplane can withstand without structural failure. If the ultimate load factor is exceeded, structural failure is imminent. 9.0 Gs.

14.   Load factor. IAS is what is the pilot sees in the cockpit.

15.   Accelerated stall lines represent the maximum load factor that an airplane can produce based on airspeed. They are also called the lines of maximum lift (C_{Lmax} AOA).

16.   Maneuvering speed is the lowest airspeed that the airplane can produce its limit load factor. T-34C V_a = 135 KIAS (at maximum gross weight).

17.   Maximum allowable speed. Redline airspeed is dependent upon M_{CRIT}, airframe temperature, structural loads, and controllability.

18.   An increase in weight decreases the limit load factor and ultimate load factor, has no effect on the redline airspeed, and sweeps the accelerated stall lines to the right.

19.   An increase in altitude will decrease the redline airspeed, but have no effect on limit load factor, ultimate load factor, or accelerated stall lines.

20.   Configuration can greatly alter every component of the safe flight envelope.

21.   When encountering asymmetric loading, the pilot must adhere to the asymmetric limit load factor which is approximately two thirds of the limit load factor. Asymmetric loading will occur as a result of a rolling pullout, trapped fuel, or hung ordnance.

22.   Gust loading. In moderate turbulence, pilot induced loading should be reduced to two thirds of the normal limit load factor, and any airspeed limitations must be strictly adhered to

23.   195 KIAS; 135 KIAS (maneuver speed)

24.   Increasing, increase

25.   increase, decrease

26.   The pilot must simultaneously move the control stick to left to begin rolling, add left rudder to overcome adverse yaw, pull aft stick to increase total lift, and increase throttle to overcome induced drag. Laterally neutralize stick approaching 30 degrees AOB.

27.   70 seconds.

28.   outside, increase, decrease

29.   skid, decrease, increase, opposite sides

TOPIC 12:

1.

	V_{TO}	V_{LDG}	IAS_{TO}	IAS_{LDG}
Weight ↑	↑	↑	↑	↑
Altitude ↑	↑	↑	N/E	N/E
Headwind	N/E	N/E	N/E	N/E
High Lift Devices	↓	↓	↓	↓

2.   Thrust, drag, lift, weight, and rolling friction. Thrust is the most unbalanced.

3.   Rolling friction is produced by the effects of friction between the tire and the runway during takeoff or landing. It is determined by multiplying the coefficient of friction by weight on wheels. It is greatest at the beginning of the takeoff roll, or the end of the landing rollout.

4.   Weight, density, wing area, C_{Lmax}, thrust, drag and rolling friction.

5.   Weight.

6.   Takeoff distance increases.

7.   Humidity increases landing distance.

8.   increase, remain the same, increase

9.   The 4-H Club (high, hot, humid, heavy) and a tailwind.

10. aerodynamic

11.

Runway	Headwind	Crosswind	Full Flap	No Flap
A	25	0	X	X
B	21.5	12.5	X	X
C	12.5	21.5		X
D	0	25		

12. Ground effect does not affect total lift. Ground effect increases effective lift, reduces induced drag, total drag, and thrust required.

13. More than 0.1, hydroplaning.

14. Wingtip vortices.

15. 2 wingspans, 1 wingspan, heavy, slow, clean

16. two

17. When landing behind a large airplane, stay at or above the other airplane's flight path and land beyond its nosewheel touchdown point. When landing behind a departing large airplane, land well prior to its rotation point.

18. Small airplanes should avoid operating within three rotor diameters of any hovering helicopter.

19. Roll clouds, gusty conditions and blowing dust.

20. Delay takeoff until sheer condition no longer exists.

21. Low level wind shear alert system (LLWAS), NEXRAD doppler radar system. PIREPS (pilot reports), departure or arrival weather reports.

Useful Equations

$$W = F \cdot s$$

$$F = ma$$

$$M = F \times d$$

$$TE = PE + KE$$

$$PE = mgh$$

$$WL = \frac{W}{S}$$

$$KE = \tfrac{1}{2} mV^2$$

$$P = \rho RT$$

$$S = bc$$

$$\lambda = \frac{C_T}{C_R}$$

$$AR = \frac{b}{c}$$

$$\rho_1 A_1 V_1 = \rho_2 A_2 V_2$$

$$P_T = P_S + q$$

$$q = \tfrac{1}{2} \rho V^2$$

$$M = \frac{TAS}{LSOS}$$

$$TAS = IAS \sqrt{\frac{\rho_0}{\rho}}$$

$$AF = \tfrac{1}{2} \rho V^2 S C_{AF}$$

$$L = \tfrac{1}{2} \rho V^2 S C_L$$

$$V_S = \sqrt{\frac{2W}{\rho S C_{Lmax}}}$$

$$IAS_S = \sqrt{\frac{2W}{\rho_0 S C_{Lmax}}}$$

$$D = \tfrac{1}{2} \rho V^2 S C_D$$

$$D_T = D_P + D_I$$

$$D_P = qf$$

$$D_I = \frac{kL^2}{\rho V^2 b^2}$$

$$P_R = \frac{T_R \cdot V}{325}$$

$$P_A = \frac{T_A \cdot V}{325}$$

$$THP = SHP \cdot p.e.$$

$$T_E = T_A - T_R$$

$$P_E = P_A - P_R$$

$$\sin \gamma = \frac{T_E}{W}$$

$$ROC = \frac{P_E}{W}$$

$$ROD = \frac{P_D}{W}$$

$$n = \frac{L}{W}$$

$$V_{S\phi} = \sqrt{\frac{2Wn}{\rho S C_{Lmax}}}$$

$$IAS_{S\phi} = \sqrt{\frac{2Wn}{\rho_0 S C_{Lmax}}}$$

$$\omega = \frac{g \tan \phi}{V}$$

$$r = \frac{V^2}{g \tan \phi}$$

$$V_{TO} \approx 1.2 \sqrt{\frac{2W}{\rho S C_{Lmax}}}$$

$$IAS_{TO} \approx 1.2 \sqrt{\frac{2W}{\rho_0 S C_{Lmax}}}$$

$$V_{LDG} \approx 1.3 \sqrt{\frac{2W}{\rho S C_{Lmax}}}$$

$$IAS_{LDG} \approx 1.3 \sqrt{\frac{2W}{\rho_0 S C_{Lmax}}}$$

$$F_R = \mu(W - L)$$

$$S_{TO} = \frac{W^2}{g \rho S C_{Lmax} (T - D - F_R)}$$

$$S_{LDG} = \frac{W^2}{g \rho S C_{Lmax} (F_R + D - T)}$$

$$V_{hydroplane} = 9 \cdot \sqrt{tire\ pressure}$$

Standard Day Conditions

Altitude (feet)	Temperature °C	Temperature °F	LSOS (knots)	Pressure (inHg)	Density (g / L)
0	15.0	59.0	661.7	29.921	1.225
1,000	13.0	55.4	659.4	28.856	1.190
2,000	11.0	51.9	657.1	27.821	1.155
3,000	9.1	48.3	654.8	26.817	1.121
4,000	7.1	44.7	652.5	25.843	1.088
5,000	5.1	41.2	650.2	24.897	1.056
6,000	3.1	37.6	647.9	23.980	1.024
7,000	1.1	34.0	645.6	23.090	0.993
8,000	-0.8	30.5	643.2	22.228	0.963
9,000	-2.8	26.9	640.9	21.391	0.934
10,000	-4.8	23.4	638.5	20.581	0.905
11,000	-6.8	19.8	636.2	19.795	0.877
12,000	-8.8	16.2	633.8	19.035	0.849
13,000	-10.7	12.7	631.4	18.298	0.823
14,000	-12.7	9.1	629.1	17.584	0.797
15,000	-14.7	5.5	626.7	16.893	0.771
16,000	-16.7	2.0	624.3	16.225	0.746
17,000	-18.7	-1.6	621.8	15.578	0.722
18,000	-20.6	-5.1	619.4	14.952	0.698
19,000	-22.6	-8.7	617.0	14.346	0.676
20,000	-24.6	-12.3	614.6	13.761	0.653
25,000	-34.5	-30.0	602.2	11.118	0.550
30,000	-44.4	-47.8	589.6	8.903	0.459
35,000	-54.2	-65.6	576.8	7.060	0.380
36,000	-56.2	-69.2	574.1	6.732	0.366
40,000	-56.5	-69.7	573.7	5.558	0.303
45,000	-56.5	-69.7	573.7	4.375	0.238
50,000	-56.5	-69.7	573.7	3.444	0.188
55,000	-56.5	-69.7	573.7	2.712	0.148
60,000	-56.5	-69.7	573.7	2.135	0.116
65,000	-56.5	-69.7	573.7	1.682	0.092

T-34C Data

PHYSICAL CHARASTERISTICS

FUSELAGE:

Construction: Semi-monocoque, Length: 28 ft 8 in, Height: 9 ft 11 in at tail

LANDING GEAR:

Tricycle

WING:

Construction: Full cantilever, Wingspan: 33 ft 5 in, Aspect ratio: 6.2:1, Loading: 24.5 lbs/ft^2

Dihedral angle: 7°, Angle of incidence: 4° at root, 0.9° at tip, Flap type: Slotted

PROPELLER:

Type: Variable pitch, Rotation: clockwise, Prop arc: 7 ft 6 in, Prop clearance: 11 in

CONTROLS:

Type: conventional – reversible

	Aerodynamic Balance	Artificial Feel
Aileron	Overhang	Servo trim tabs
Rudder	Shielded Horn	Anti-servo trim tabs
Elevator	Shielded Horn	Bobweights and Downsprings

FLIGHT CHARACTERISTICS

C_{Lmax} AOA: 29 to 29.5 units

Stall warning: Buffet, rudder shakers, AOA indicator, AOA indexer

Spin indications:

	Erect	Inverted
Altimeter	Decreasing	Decreasing
AOA	30 Units	2–3 Units
Airspeed	80–100 KIAS	Zero
Turn Needle	Pegged	Pegged

Crosswind limits: 22 kts (no flap), 15 kts (full flap)

Redline airspeed: 280 KIAS, 245 KIAS (>20,000' MSL), 150 KIAS (gear down), 120 KIAS (flaps down)

Maneuver airspeed: 135 KIAS, Maximum turbulence airspeed: 195 KIAS

Limit load factor: +4.5 Gs, −2.3 Gs; Flaps down: +2.0 Gs, −1.0 G

Standard rate turn: 2 needlewidths; AOB = 15–20% indicated airspeed

PERFORMANCE CHARACTERISTICS

Engine: PT6A-25, Sea level flat rated 550 SHP (1315 ft-lbs); Navy limited to 425 SHP (1015 ft-lbs)

Critical altitude: 14,000 ft

Operational ceiling: 25,000 ft

Max endurance achieved at 420 ft-lbs torque, max range at 580 ft-lbs torque

Max AOC airspeed: 75 KIAS, max ROC airspeed: 100 KIAS

Max level airspeed: 190 KIAS, normal climb airspeed: 120 KIAS

Min sink rate achieved at 87 KIAS, Max glide range at 98 KIAS (operational glide airspeed: 100 KIAS)

CONTRIBUTIONS TO STATIC STABILITY

Feature	Longitudinal	Directional	Lateral
Straight Wings	-	+	
Swept / Delta Wings	+	+	+
Fuselage	-	-	
Horizontal Stabilizer	++		
Neutral Point. aft of C.G.	+		
Vertical Stabilizer		++	+
Dihedral Wings			++
Anhedral Wings			--
High-Mounted Wings			+
Low Mounted Wings			-

Performance Study Guide

	Weight Increase	Altitude Increase	Gears Down	Flaps Down
T_A		↓		
T_R	→↑	→	↑	←↑
T_E	↓	↓	↓	↓
P_A		↓		
P_R	→↑	→↑	↑	←↑
P_E	↓	↓	↓	↓

Performance Characteristic	Goal	Curve Referenced & V vs L/D$_{MAX}$		Effect of Factors on Performance			
		Jet	Prop	Weight↑	Altitude↑	Tailwind↑	Gear↓/Flap↓
Endurance	Min Fuel Flow	Thrust, =	Power, <	↓	↑		↓
Range	Min Fuel Flow / Velocity	Thrust, >	Power, =	↓	↑	↑	↓
AOC	Max T_E	Thrust, =	Thrust, <	↓	↓	↓	↓
ROC	Max P_E	Power, >	Power, =	↓	↓		↓
Glide Endurance	Min P_D	Power, <	Power, <	↓	↑		↓
Glide Range	Min T_D	Thrust, =	Thrust, =		↑	↑	↓

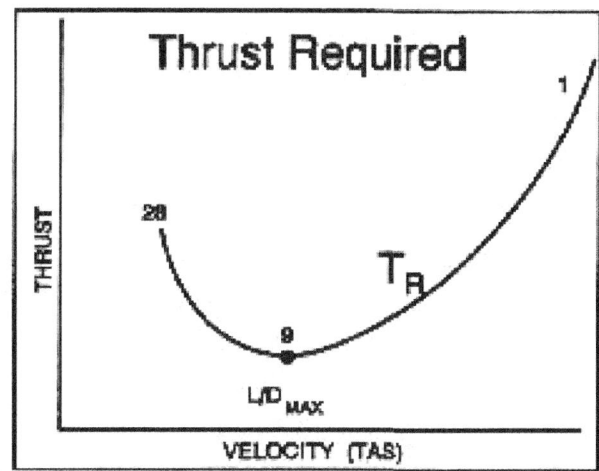

Figure 1

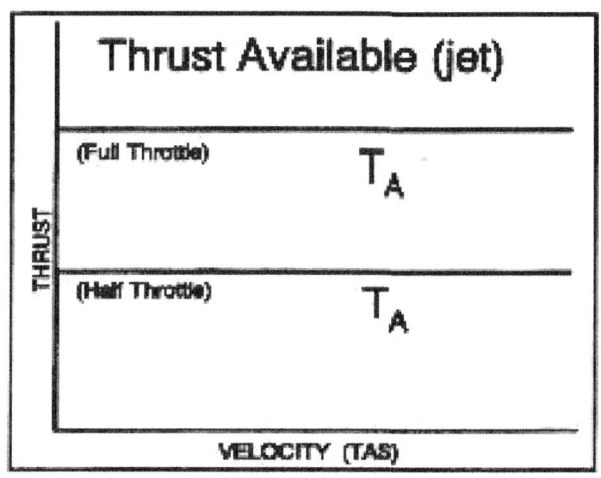

Figure 2

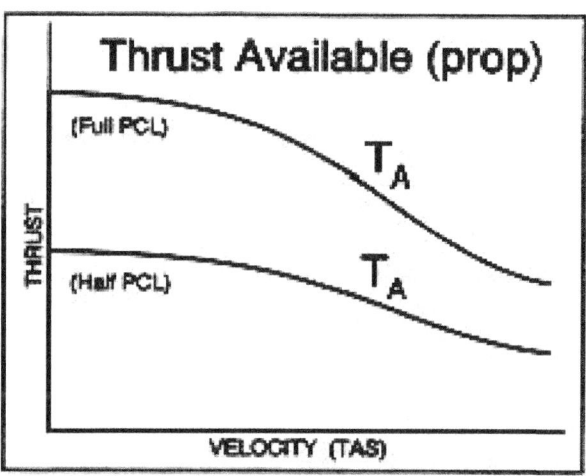

Figure 3

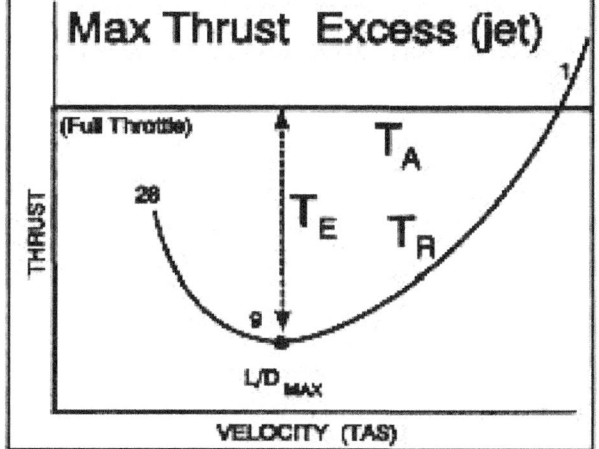

Figure 4

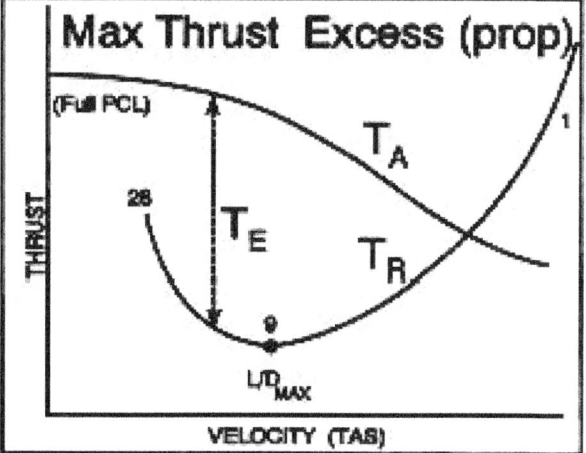

Figure 5

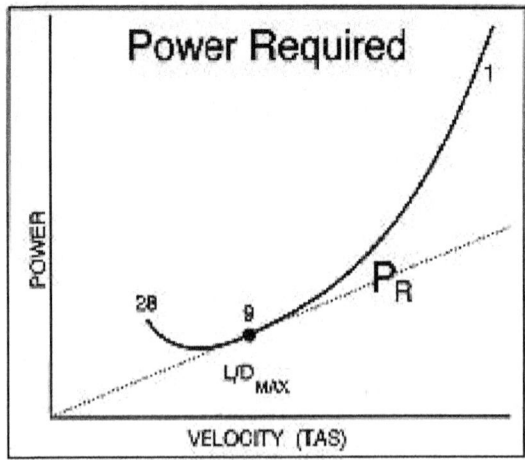

Figure 6

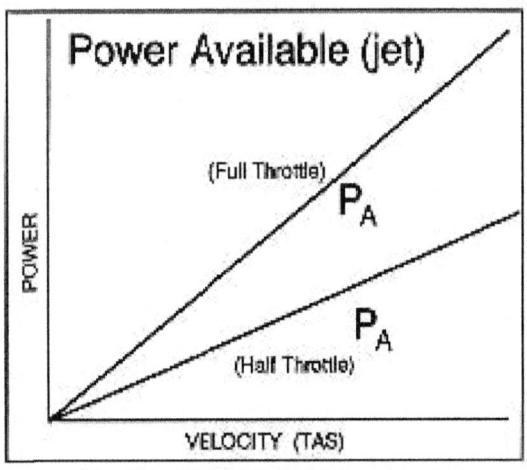

Figure 7

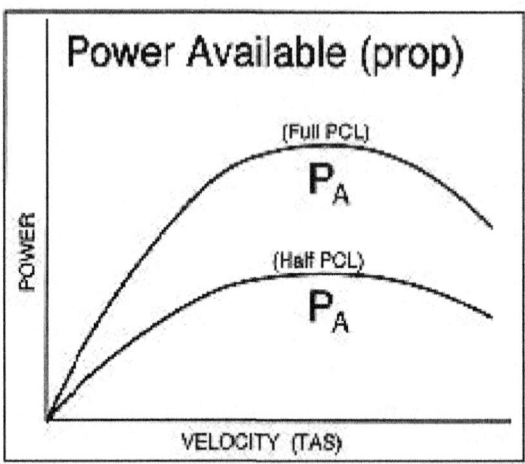

Figure 8

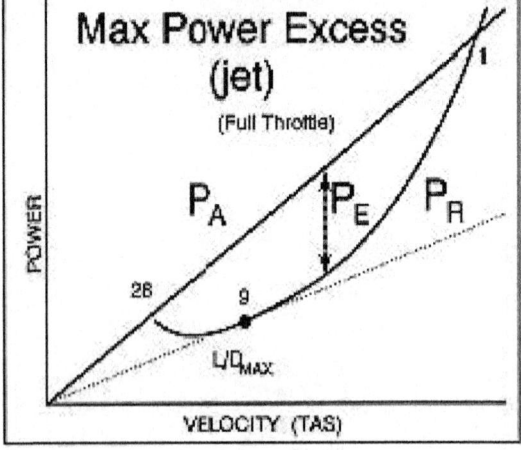

Figure 9

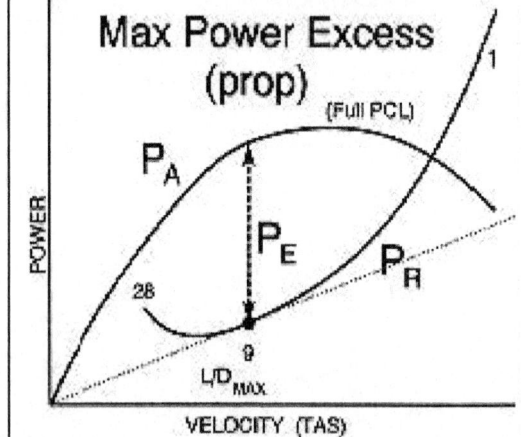

Figure 10

Symbols & Abbreviations

α	angle of attack	C_D	coefficient of drag
β	sideslip angle	C_{Di}	coefficient of induced drag
γ	climb or glide angle	C_{Dp}	coefficient of parasite drag
θ	pitch attitude	C_F	coefficient of aerodynamic force
λ	taper ratio	CG	center of gravity
Λ	sweep angle	C_L	coefficient of lift
μ	viscosity or coefficient of friction	C_{LMAX}	maximum coefficient of lift
ρ	air density	c_R	root chord
ϕ	angle of bank	c_T	tip chord
ω	rate of turn	d	moment arm distance
<	less than	D	drag
>	greater than	D	Adensity altitude
$\approx$	approximately equal to	D_I	induced drag
a	acceleration	D_p	parasite drag
A	area	D_T	total drag
AC	aerodynamic center	EAS	equivalent airspeed
AF	aerodynamic force	f	equivalent parasite area
AOA	angle of attack	F	force
AOC	angle of climb	FF	fuel flow
AR	aspect ratio	F_R	rolling friction
b	wing span	G	acceleration of gravity
c	average chord	GS	groundspeed
CAS	calibrated airspeed	IAS	indicated airspeed

IAS_{LDG} indicated landing speed

IAS_S indicated stall speed

$IAS_{S\phi}$ indicated accelerated stall speed

IAS_{SP} indicated power-off stall speed

IAS_{TO} indicated takeoff speed

k constant

KE kinetic energy

L/D lift to drag ratio

L/D_{MAX} maximum lift to drag ratio

L lift

L_{EFF} effective lift

LSOS local speed of sound

m mass

M Mach number

M_{CRIT} critical Mach number

n load factor

NP neutral point

NWLO nosewheel liftoff speed

NWTD nosewheel touchdown speed

P pressure

p.e. propeller efficiency

PE potential energy

PA pressure altitude

P_A power available

P_D power deficit

P_E power excess

P_R power required

P_S static pressure

P_T total pressure

q dynamic pressure

Q torque

r radius of turn

ROC rate of climb

RW relative wind

s distance of displacement

s_{LDG} landing distance

s_{TO} takeoff distance

S wing surface area

SHP shaft horsepower

SRT standard rate turn

T temperature or thrust

T_A thrust available

T_D thrust deficit

Bibliography

Comeaux, J. Aerodynamics for Pilots (ATC Pamphlet 51-3), Washington, D.C.: U.S. Government Printing Office, 1979.

Dommasch, Daniel O.; Sherby, Sydney S.; and Connolly, Thomas F. Airplane Aerodynamics, 4th Ed., New York: Pitman Publishing Corporation, 1967.

Etkin, Bernard. Dynamics of Flight-Stability and Control, New York: Wiley, 1982.

Hurt, Hugh. Aerodynamics for Naval Aviators (NAVWEPS 00-80T-80). Washington, D.C.: U.S. Government Printing Office, 1960.

Kuethe, Arnold M.; and Chow, Chuen-Yen. Foundations of Aerodynamics, New York: Wiley, 1976.

Shevell, Richard S. Fundamentals of Flight, Englewood Cliffs, NJ: Prentice-Hall, Inc., 1983.

Smith, Hubert C. The Illustrated Guide to Aerodynamics, 2nd Ed., Blue Ridge Summit, PA: TAB Books, 1992.

Office of the Chief of Naval Operations. T-34C NATOPS Flight Manual (NAVAIR 01-T34AAC-1), Commander Naval Air Systems Command, 1988.

Talay, Theodore A. Introduction to the Aerodynamics of Flight (NASA SP-367), Langley Research Center, 1975.

National Oceanic and Atmospheric Administration. U.S. Standard Atmosphere, 1976, Washington, D.C.: U.S. Government Printing Office, 1976.

Change Recommendation

TO BE FILLED IN BY ORIGINATOR AND FORWARDED TO MODEL MANAGER					
FROM (originator)		Date			
TO (Model Manager) Aerodynamics Branch Officer		NAVAVSCOLSCOM			
Name of Student Guide	Revision Date	Change Date	Chapter	Page	Paragraph
Fundamentals of Aerodynamics	March 2008				

Recommendation (be specific)

CONTINUE ON SEPARATE SHEET IF NEEDED

Justification (indicate references if needed)

SIGNATURE	Rank	Title

Address of Unit or Command

TO BE FILLED IN BY MODEL MANAGER (Return to Originator)	
FROM	DATE
TO	

Your change recommendation dated _____ is acknowledged. It will be held for action of the review conference planned for _____ to be held at NAVAVSCOLSCOM.

MODEL MANAGER

Printed in Great Britain
by Amazon

47366073R00130